MATH GAMES
Skill-Based Practice for First Grade

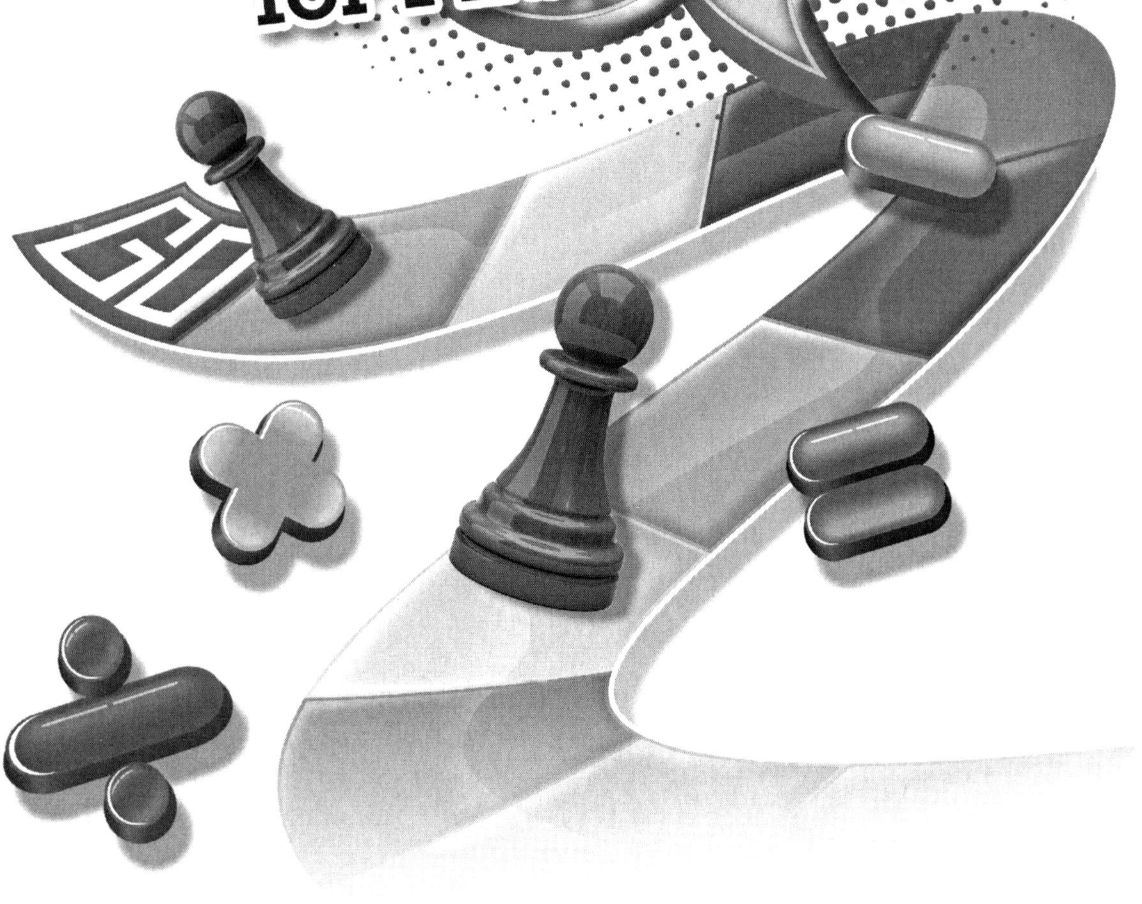

Authors
Ted H. Hull, Ed.D.
Ruth Harbin Miles, Ed.S.
Don S. Balka, Ph.D.

Consultant

Don W. Scheuer, Jr., M.S.Ed.
Mathematics Specialist
The Haverford School (ret.)

Publishing Credits

Robin Erickson, *Production Director;*
Lee Aucoin, *Creative Director;* Tim J. Bradley, *Illustration Manager;*
Sara Johnson, M.S.Ed., *Editorial Director;* Maribel Rendón M.A.Ed., *Editor;*
Jennifer Viñas, *Editor;* Grace Alba, *Designer;*
Corinne Burton, M.A.Ed., *Publisher*

Image Credits

all images Shutterstock

Standards

© 2007 Teachers of English to Speakers of Other Languages, Inc. (TESOL)
© 2007 Board of Regents of the University of Wisconsin System. World-Class Instructional Design and Assessment (WIDA). For more information on using the WIDA ELP Standards, please visit the WIDA website at www.wida.us.
© 2010 National Governors Association Center for Best Practices and Council of Chief State School Officers (CCSS)

Shell Education
5301 Oceanus Drive
Huntington Beach, CA 92649-1030
http://www.shelleducation.com
ISBN 978-1-4258-1288-1
© 2014 Shell Educational Publishing, Inc.

The classroom teacher may reproduce copies of materials in this book for classroom use only. The reproduction of any part for an entire school or school system is strictly prohibited. No part of this publication may be transmitted, stored, or recorded in any form without written permission from the publisher.

Table of Contents

Introduction

    Importance of Games 5

    Mathematical Learning 5

    Games vs. Worksheets 6

How to Use This Book 8

Correlation to the Standards 11

About the Authors 15

Math Games

Domain: Operations and Algebraic Thinking

    Post Office Game 16

    I Can Solve It! 38

    Mix-Up, Match-Up 42

    It's a Fact! 47

    Speedy Counters 51

    Climb to the Top 60

    True or False 64

    Unknown Number Game 67

Domain: Numbers and Operations in Base Ten

    The Dice Game 72

    Place Value 73

    Roll a Place Value Game 77

    Handy Math 80

    Ten: More or Less 87

    Multiples Subtraction 91

Table of Contents (cont.)

Domain: Measurement and Data

    Simon Says, Compare Me 94

    Measurement, Measurement, Measurement 96

    It's the Right Time 100

    Graph It! 105

Domain: Geometry

    Tell Me About It Geometry 121

    Shape-Maker 125

    Sharing Shapes 131

Appendices

Appendix A: References Cited 138

Appendix B: Contents of the Digital Resource CD 139

Importance of Games

Students learn from play. Play begins when we are infants and continues through adulthood. Games are motivational and educational (Hull, Harbin Miles, and Balka 2013; Burns 2009). They can assist and encourage students to operate as learning communities by requiring students to work together by following rules and being respectful. Games also foster students' thinking and reasoning since students formulate winning strategies. They provide much more sustained practices than do worksheets, and students are more motivated to be accurate. Worksheets may provide 20 to 30 opportunities for students to practice a skill, while games far exceed such prescribed practice opportunities. Lastly, games provide immediate feedback to students concerning their abilities.

Games must be part of the overall instructional approach that teachers use because successful learning requires active student engagement (Hull, Harbin Miles, and Balka 2013; National Research Council 2004), and games provide students with the motivation and interest to become highly engaged. Instructional routines need balance between concept development and skill development. They must also balance teacher-led and teacher-facilitated lessons. Students need time to work independently and collaboratively in order to assimilate information, and games can help support this.

When games are used appropriately, students also learn mathematical concepts.

Mathematical Learning

Students must learn mathematics with understanding (NCTM 2000). Understanding means that students know the relationship between mathematical concepts and mathematical skills—mathematical procedures and algorithms work because of the underlying mathematical concepts. In addition, skill proficiency allows students to explore more rigorous mathematical concepts. From this relationship, it is clear that a balance between skill development and conceptual development must exist. There cannot be an emphasis of one over the other.

The National Council of Teachers of Mathematics (2000) and the National Research Council (2001) reinforce this idea. Both organizations state that learning mathematics requires both conceptual understanding and procedural fluency. This means that students need to practice procedures as well as develop their understanding of mathematical concepts in order to achieve success. The games presented in this book reinforce skill-based practice and support students' development of proficiency. These games can also be used as a springboard for discourse about mathematical concepts. The counterpart to this resource is *Math Games: Getting to the Core of Conceptual Understanding*, which builds students' conceptual understanding of mathematics through games.

Importance of Games (cont.)

The *Common Core State Standards for Mathematics* (2010) advocate a balanced mathematics curriculum by focusing standards both on mathematical concepts and skills. This is also stressed in the Standards for Mathematical Practice, which discuss the process of "doing" mathematics and the habits of mind students need to possess in order to be successful.

The Standards for Mathematical Practice also focus on the activities that foster thinking and reasoning in which students need to be involved while learning mathematics. Games are an easy way to initiate students in the development of many of the practices. Each game clearly identifies a Common Core domain, a standard, and a skill, and allows students to practice them in a fun and meaningful way.

Games vs. Worksheets

In all likelihood, many mathematics lessons are skill related and are taught and practiced through worksheets. Worksheets heavily dominate elementary mathematics instruction. They are not without value, but they often command too much time in instruction. While students need to practice skills and procedures, the way to practice these skills should be broadened.

Worksheets generally don't promote thinking and reasoning. They become so mechanical that students cease thinking. They are lulled into a feeling that completing is the goal. This sense of "just completing" is not what the Common Core Standards for Mathematical Practice mean when they encourage students to "persevere in solving problems."

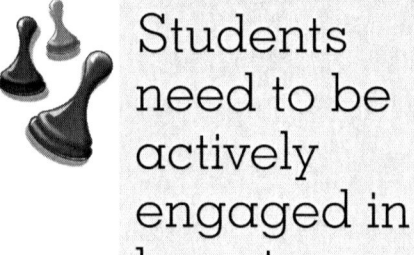

Students need to be actively engaged in learning.

Students need to be actively engaged in learning. While worksheets do serve a limited purpose in skill practice, they also contain many potential difficulties. Problems that can occur include the following:

➜ **Worksheets are often completed in isolation,** meaning that students who are performing a skill incorrectly most likely practice the skill incorrectly for the entire worksheet. The misunderstanding may not be immediately discovered, and in fact, will most likely not be discovered for several days!

➜ **Worksheets are often boring to students.** Learning a skill correctly is not the students' goal. Their goal becomes to finish the worksheet. As a result, careless errors are often made, and again, these errors may not be immediately discovered or corrected.

Importance of Games (cont.)

→ **Worksheets are often viewed as a form of subtle punishment.** While perhaps not obvious, the perceived punishment is there. Students who have mastered the skill and can complete the worksheet correctly are frequently "rewarded" for their efforts with another worksheet while they wait for their classmates to finish. At the same time, students who have not mastered the skill and do not finish the worksheet on time are "rewarded" with the requirement to take the worksheet home to complete, or they finish during another portion of the day, often recess or lunch.

→ **Worksheets provide little motivation to learn a skill correctly.** There is no immediate correction for mistakes, and often, students do not really care if a mistake is made. When a game is involved, students want and need to get correct answers.

The *Common Core State Standards for Mathematics*, including the Standards for Mathematical Practice, demand this approach change. These are the reasons teachers and teacher leaders must consciously support the idea of using games to support skill development in mathematics.

How to Use This Book

There are many ways to effectively utilize this book. Teachers, mathematics leaders, and parents may use this book to engage students in fun, meaningful, practical mathematics learning. These games can be used as a way to help students maintain skill proficiency or remind them of particular skills prior to a critical concept lesson. These games may also be useful during tutorial sessions, or during class when students have completed their work.

Games at Home

Parents may use these games to work with their child to learn important skills. The games also provide easier ways for parents to interest their child in learning mathematics rather than simply memorizing facts. In many cases, their child is more interested in listening to explanations than correcting their errors.

Parents want to help their children succeed in school, yet they may dread the frequently unpleasant encounters created by completing mathematics worksheets at home. Families can easily use the games in this book by assuming the role of one of the players. At other times, parents provide support and encouragement as their child engages in the game. In either situation, parents are able to work with their children in a way that is fun, educational, and informative.

Games in the Classroom

During game play, teachers are provided excellent opportunities to assess students' abilities and current skill development. Students are normally doing their best and drawing upon their current understanding and ability to play the games, so teachers see an accurate picture of student learning. Some monitoring ideas for teacher assessment include:

- ➔ Move about the room listening and observing
- ➔ Ask student pairs to explain what they are doing
- ➔ Ask the entire class about the game procedures after play
- ➔ Play the game against the class
- ➔ Draw a small group of students together for closer supervision
- ➔ Gather game sheets to analyze students' proficiencies

Ongoing formative assessment and timely intervention are cornerstones of effective classroom instruction. Teachers need to use every available opportunity to make student thinking visible and to respond wisely to what students' visible thinking reveals. Games are an invaluable instructional tool that teachers need to effectively use.

How to Use This Book (cont.)

Students are able to work collaboratively during game play, thus promoting student discourse and deeper learning. The games can also be used to reduce the amount of time students spend completing worksheets.

Each game in this book is based upon a common format. This format is designed to assist teachers in understanding how the game activities are played and which standards and mathematical skills students will be practicing.

Domain
The domain that students will practice is noted at the beginning of each lesson. Each of the four domains addressed in this series has its own icon.

Standards
One or more *Common Core State Standards* will state the specific skills that students will practice during game play.

Number of Players
The number of players varies for each game. Some may include whole-group game play, while others may call for different-size groups.

Materials
A materials list is provided for each game to notify the teacher what to have available in order to play the games.

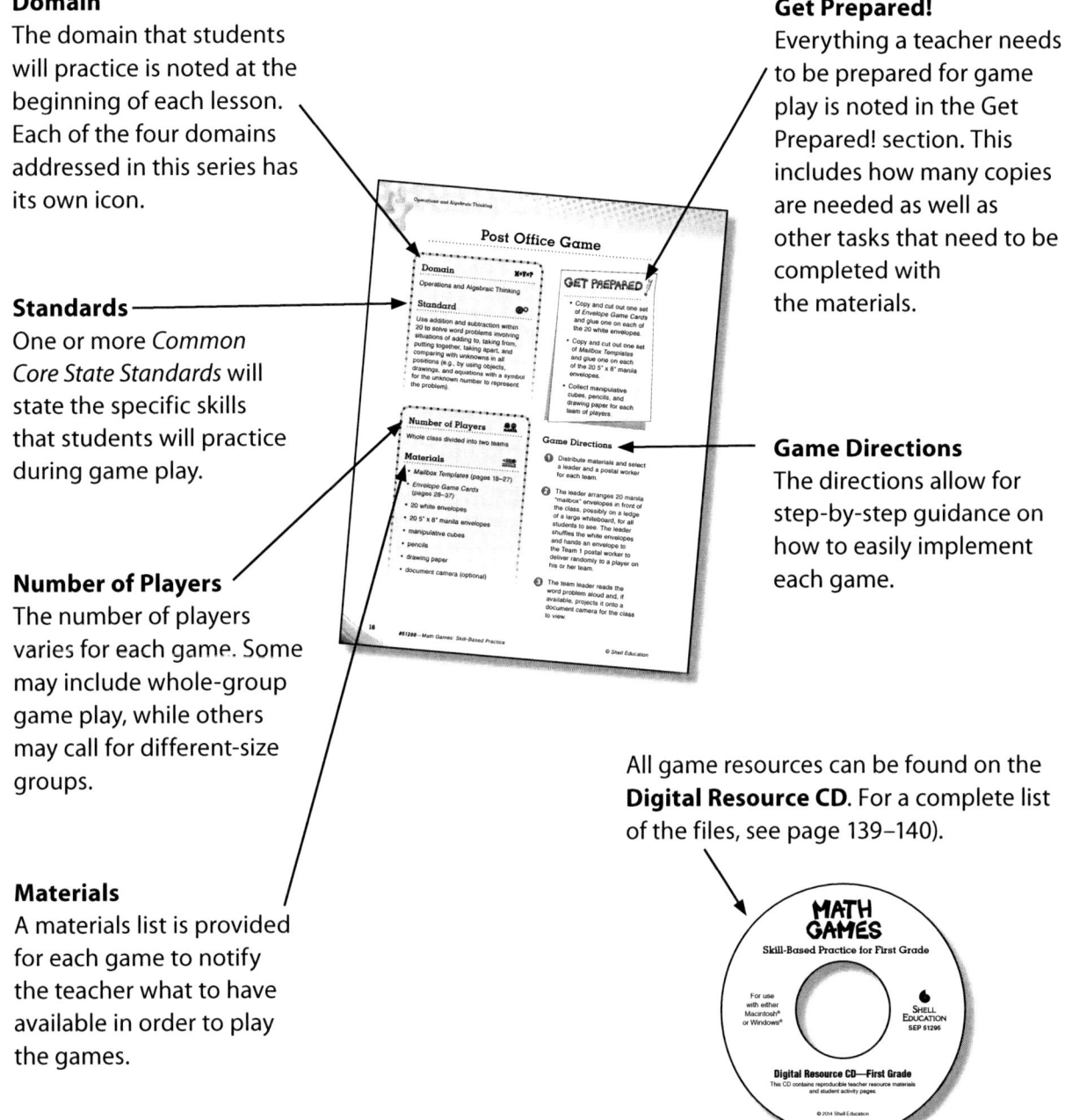

Get Prepared!
Everything a teacher needs to be prepared for game play is noted in the Get Prepared! section. This includes how many copies are needed as well as other tasks that need to be completed with the materials.

Game Directions
The directions allow for step-by-step guidance on how to easily implement each game.

All game resources can be found on the **Digital Resource CD**. For a complete list of the files, see page 139–140).

How to Use This Book (cont.)

Many games include materials such as game boards, activity cards, score cards, and spinners. You may wish to laminate materials for durability.

Game Boards
Some game boards spread across multiple book pages in order to make them larger for game play. When this is the case, cut out each part of the game board and tape them together. Once you cut them apart and tape them together, you may wish to glue them to a large sheet of construction paper and laminate them for durability.

Activity Cards
Some games include activity cards. Once you cut them apart, you may wish to laminate them for durability.

Spinners
Some games include spinners. To use a spinner, cut it out from the page. Place the tip of a pencil in the center with a paperclip around it. Use your other hand to flick the other side of the paperclip.

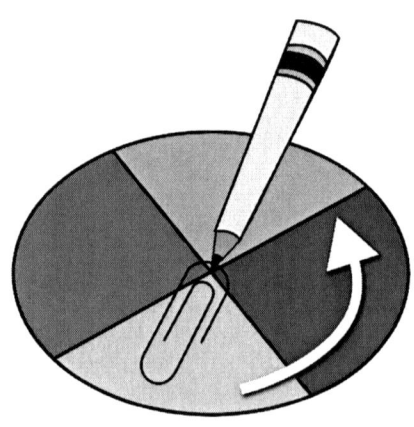

Correlation to the Standards

Shell Education is committed to producing educational materials that are research and standards based. In this effort, we have correlated all of our products to the academic standards of all 50 United States, the District of Columbia, the Department of Defense Dependent Schools, and all Canadian provinces.

How To Find Standards Correlations

To print a customized correlation report of this product for your state, visit our website at **http://www.shelleducation.com** and follow the on-screen directions. If you require assistance in printing correlation reports, please contact Customer Service at 1-877-777-3450.

Purpose and Intent of Standards

Legislation mandates that all states adopt academic standards that identify the skills students will learn in kindergarten through grade twelve. Many states also have standards for Pre–K. This same legislation sets requirements to ensure the standards are detailed and comprehensive.

Standards are designed to focus instruction and guide adoption of curricula. Standards are statements that describe the criteria necessary for students to meet specific academic goals. They define the knowledge, skills, and content students should acquire at each level. Standards are also used to develop standardized tests to evaluate students' academic progress. Teachers are required to demonstrate how their lessons meet state standards. State standards are used in the development of all of our products, so educators can be assured they meet the academic requirements of each state.

Common Core State Standards

Many games in this book are aligned to the Common Core State Standards. The standards support the objectives presented throughout the lessons and are provided on the Digital Resource CD (standards.pdf).

TESOL and WIDA Standards

The lessons in this book promote English language development for English language learners. The standards listed on the Digital Resource CD (standards.pdf) support the language objectives presented throughout the lessons.

Correlation to the Standards

Standards Correlation Chart

Standard	Game(s)
1.OA.1—Use addition and subtraction within 20 to solve word problems involving situations of adding to, taking from, putting together, taking apart, and comparing, with unknowns in all positions e.g., by using objects, drawings, and equations with a symbol for the unknown number to represent the problem.	Post Office Game (p. 16)
1.OA.A.2—Solve word problems that call for addition of three whole numbers whose sum is less than or equal to 20, e.g., by using objects, drawings, and equations with a symbol for the unknown number to represent the problem.	I Can Solve It! (p. 38)
1.OA.B.3—Apply properties of operations and strategies to add and subtract. *Example: if 8 + 3 = 11 is known, then 3 + 8 = 11 is also known (commutative property).*	Mix-Up, Match-Up (p. 42)
1.OA.B.4—Understand subtraction as an unknown-addend problem. *For example, subtract 10 – 8 by finding the number that makes 10 when added to 8.*	It's a Fact! (p. 47)
1.OA.C.5—Relate counting to addition and subtraction (e.g., by counting on 2 to add 2).	Speedy Counters (p. 51)
1.OA.C.6—Add and subtract within 20, demonstrating fluency for addition and subtraction within 10. Use strategies such as counting on, making ten, decomposing a number leading to ten, using the relationship between addition and subtraction, and creating equivalents to easier or known sums.	Climb to the Top (p. 60)
1.OA.D.7—Understand the meaning of the equal sign, and determine if equations involving addition and subtraction are true or false. *For example, which of the following equations are true and which are false? 6 = 6, 7 = 8 – 1, 5 + 2 = 2 + 5, 4 + 1 = 5 + 2.*	True or False (p. 64)
1.OA.D.8—Determine the unknown whole number in an addition or subtraction equation relating three whole numbers. *For example, determine the unknown number that makes the equation true in each of the equations: 8 + ? = 11, 5 = ? – 3, 6 + 6 = ?*	Unknown Number Game (p. 67)
1.NBT.A.1—Count to 120 starting at any number less than 120. In this range, read and write numerals and represent a number of objects with a written numeral.	The Next 3… (p. 72)

Standards Correlation Chart (cont.)

Standard	Game(s)
1.NBT.B.2—Understand that the two digits of a two-digit number represent amounts of tens and ones.	Place Value (p. 73)
1.NBT.B.3—Compare two two-digit numbers based on the meanings of the tens and ones digits, recording the results of comparisons with the symbols >, =, and <.	Roll It! Place Value Game (p. 77)
1.NBT.C.4—Add within 100, including adding a two-digit number and a one-digit number and adding a two-digit number and a multiple of 10, using concrete models or drawings and strategies based on place value, properties of operations, and/or the relationship between addition and subtraction; relate the strategy to a written method and explain the reasoning used. Understand that in adding two-digit numbers, one adds tens and tens, and ones and ones; and sometimes it is necessary to compose a ten.	Handy Math (p. 80)
1.NBT.C.5—Given a two-digit number, mentally find 10 more or 10 less than the number without having to count; explain the reasoning used.	Ten: More or Less (p. 87)
1.NBT.C.6—Subtract multiples of 10 in the range 10–90 from multiples of 10 in the range of 10–90 (positive or zero differences), using concrete models or drawings and strategies based on place value, properties of operations and/or the relationship between addition and subtraction; relate the strategy to a written method and explain the reasoning used.	Multiples Subtraction (p. 91)
1.MD.A.1—Order three objects by length; compare the lengths of two objects indirectly by using a third object.	Simon Says, Compare Me (p. 94)
1.MD.A.2—Express the length of an object as a whole number of length units, by laying multiple copies of a shorter object (the length unit) end to end; understand that the length measurement of an object is the number of same-size length units that span it with no gaps or overlaps. *Limit to contexts where the object being measured is spanned by a whole number of length units with no gap or overlaps.*	Measurement, Measurement, Measurement (p. 96)
1.MD.B.3—Tell and write time in hours and half-hours using analog and digital clocks.	It's the Right Time (p. 100)

Correlation to the Standards

Standards Correlation Chart (cont.)

Standard	Game(s)
1.MD.C.4—Organize, represent, and interpret data with up to three categories; ask and answer questions about the total number of data points, how many in each category, and how many more or less are in one category than in another.	Graph It! (p. 105)
1.G.A.1—Distinguish between defining attributes (e.g., triangles are closed and three-sided), versus non-defining attributes (e.g., color, orientation, overall size); build and draw shapes to possess defining attributes.	Tell Me About It Geometry (p. 121)
1.G.A.2—Compose two-dimensional shapes (rectangles, squares, trapezoids, triangles, half-circles, and quarter-circles) or three-dimensional shapes (cubes, right rectangular prisms, right circular cones, and right circular cylinders) to create a composite shape, and compose new shapes from the composite shape.	Shape-Maker (p. 125)
1.G.A.3—Partition circles and rectangles into two and four equal shares, describe the shapes using the words *halves*, *fourths*, and *quarters*, and use the phrases *half of*, *fourth of*, and *quarter of*. Describe the whole as two of, or four of the shares. Understand for these examples that decomposing into more equal shares creates smaller shares.	Sharing Shapes (p. 131)

About the Authors

Ted H. Hull, Ed.D., served in public education for 32 years as a mathematics teacher, a K–12 mathematics coordinator, a school principal, director of curriculum and instruction, and project director for the Charles A. Dana Center at the University of Texas in Austin. While at the University of Texas, he directed the research project "Transforming Schools: Moving from Low-Achieving to High Performing Learning Communities." After retiring, Ted opened LCM: Leadership • Coaching • Mathematics with his coauthors and colleagues. Ted has coauthored numerous books addressing mathematics improvement and has served as the Regional Director for the National Council of Supervisors of Mathematics (NCSM).

Ruth Harbin Miles, Ed.S., currently coaches inner-city, rural, and suburban mathematics teachers and serves on the Board of Directors for the National Council of Teachers of Mathematics, the National Council of Supervisors of Mathematics and Virginia's Council of Mathematics Teachers. Her professional experiences include coordinating the K–12 Mathematics Department for Olathe, Kansas Schools and adjunct teaching for Mary Baldwin College and James Madison University in Virginia. A coauthor of four books on transforming teacher practice through team leadership, mathematics coaching, and visible student thinking and co-owner of Happy Mountain Learning, Ruth's specialty and passion include developing teachers' content knowledge and strategies for engaging students to achieve high standards in mathematics.

Don S. Balka, Ph.D., a former middle school and high school mathematics teacher, is Professor Emeritus in the Mathematics Department at Saint Mary's College in Notre Dame, Indiana. Don has presented at over 2,000 workshops, conferences, and in-service trainings throughout the United States and has authored or coauthored over 30 books on mathematics improvement. Don has served as director for the National Council of Teachers of Mathematics, the National Council of Supervisors of Mathematics, TODOS: Mathematics for All, and the School Science and Mathematics Association. He is currently president of TODOS and past president of the School Science and Mathematics Association.

Post Office Game

Domain X+Y=?

Operations and Algebraic Thinking

Standard

Use addition and subtraction within 20 to solve word problems involving situations of adding to, taking from, putting together, taking apart, and comparing with unknowns in all positions (e.g., by using objects, drawings, and equations with a symbol for the unknown number to represent the problem).

Number of Players

Whole class divided into two teams

Materials

- *Mailbox Templates* (pages 18–27)
- *Envelope Game Cards* (pages 28–37)
- 19 white envelopes
- 20 5" x 8" manila envelopes
- manipulative cubes
- pencils
- drawing paper
- document camera (optional)

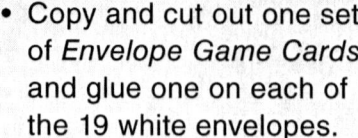

- Copy and cut out one set of *Envelope Game Cards* and glue one on each of the 19 white envelopes.
- Copy and cut out one set of *Mailbox Templates* and glue one on each of the 20 5" x 8" manila envelopes.
- Collect manipulative cubes, pencils, and drawing paper for each team of players.

Game Directions

1. Distribute materials and select a leader and a postal worker for each team.

2. The leader arranges 20 manila "mailbox" envelopes in front of the class, possibly on a ledge of a large whiteboard, for all students to see. The leader shuffles the white envelopes and hands an envelope to the Team 1 postal worker to deliver randomly to a player on his or her team.

3. The team leader reads the word problem aloud and, if available, projects it onto a document camera for the class to view.

Post Office Game (cont.)

4. The player solves the word problem and "mails" the problem by putting it in the manila envelope with the mailbox that shows the correct answer. Use of manipulative cubes and/or drawings is encouraged. The player may ask other players on the team for assistance in solving the problem; however, the player must explain his or her thinking prior to mailing the letter word problem.

5. If correct, Team 1 scores a point. If incorrect, the leader can put the letter in the "dead letter" box envelope.

6. The game continues with the postal worker on Team 2 delivering an envelope for a player on his or her team to solve.

7. After all of the problems are solved, the team with the most letters correctly "delivered" wins!

Operations and Algebraic Thinking

Mailbox
Templates

Directions: Copy and cut out one set of mailbox cards. Glue each card onto a manila envelope.

Mailbox
Templates (cont.)

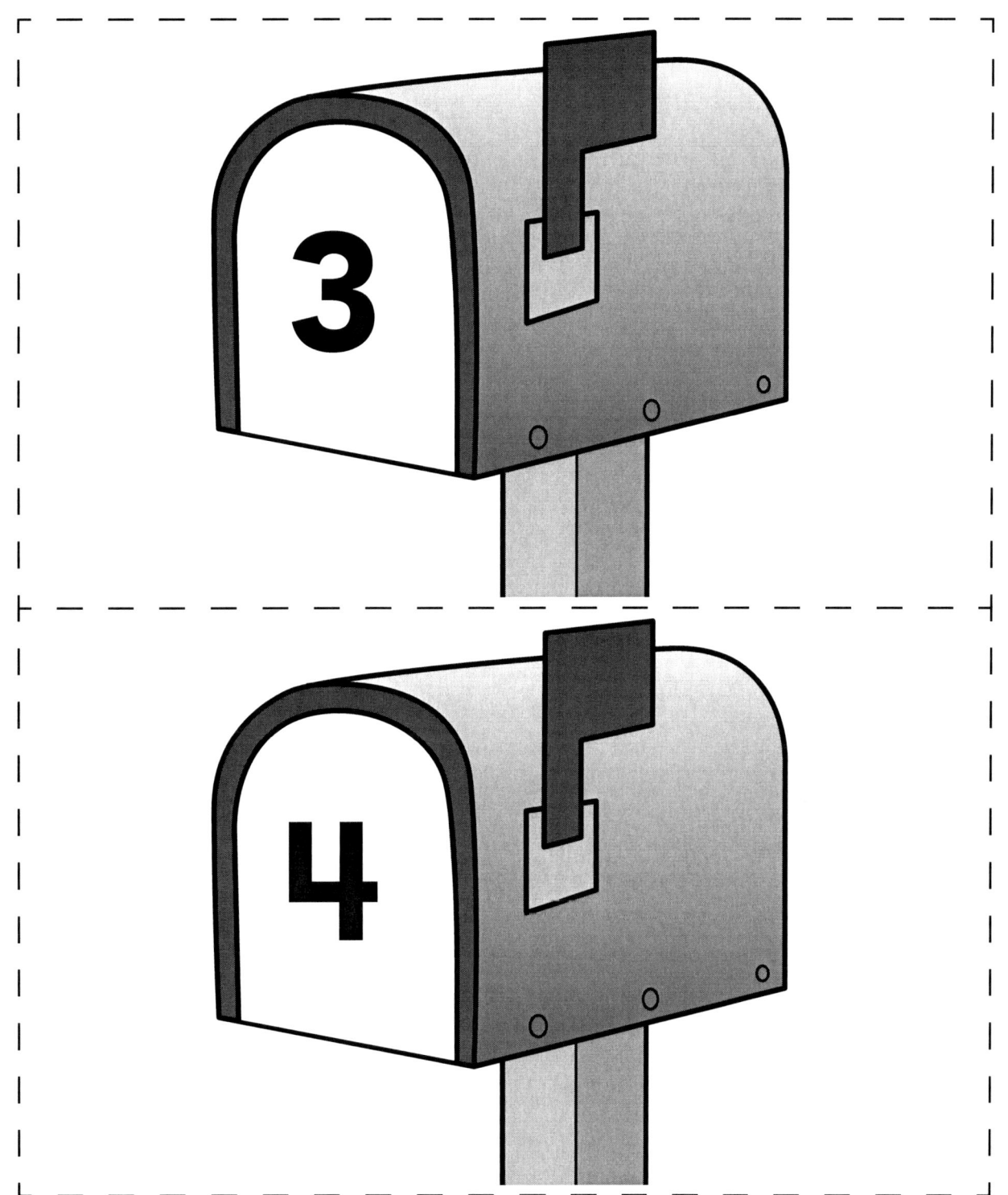

Operations and Algebraic Thinking

Mailbox
Templates (cont.)

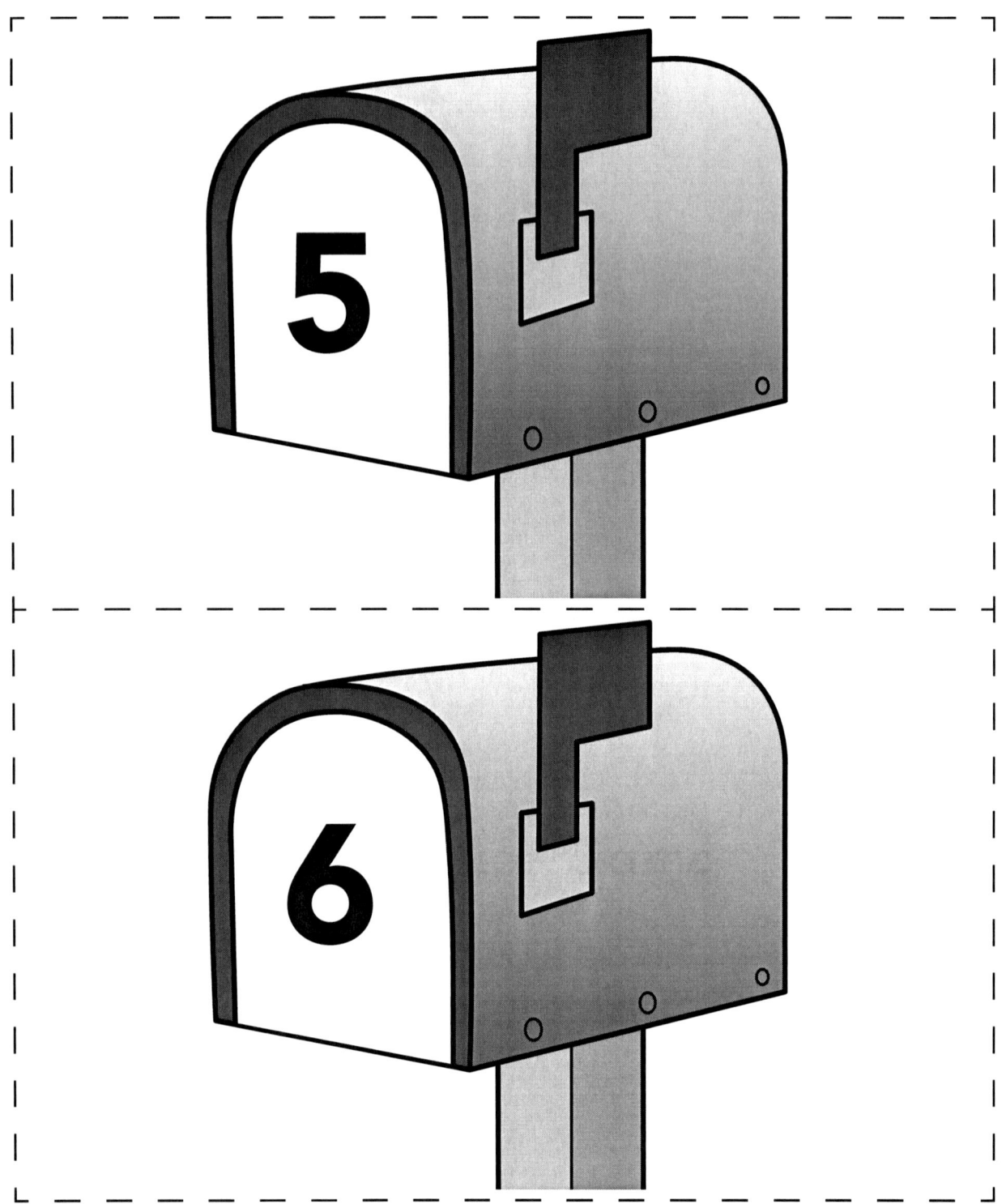

Operations and Algebraic Thinking

Mailbox
Templates (cont.)

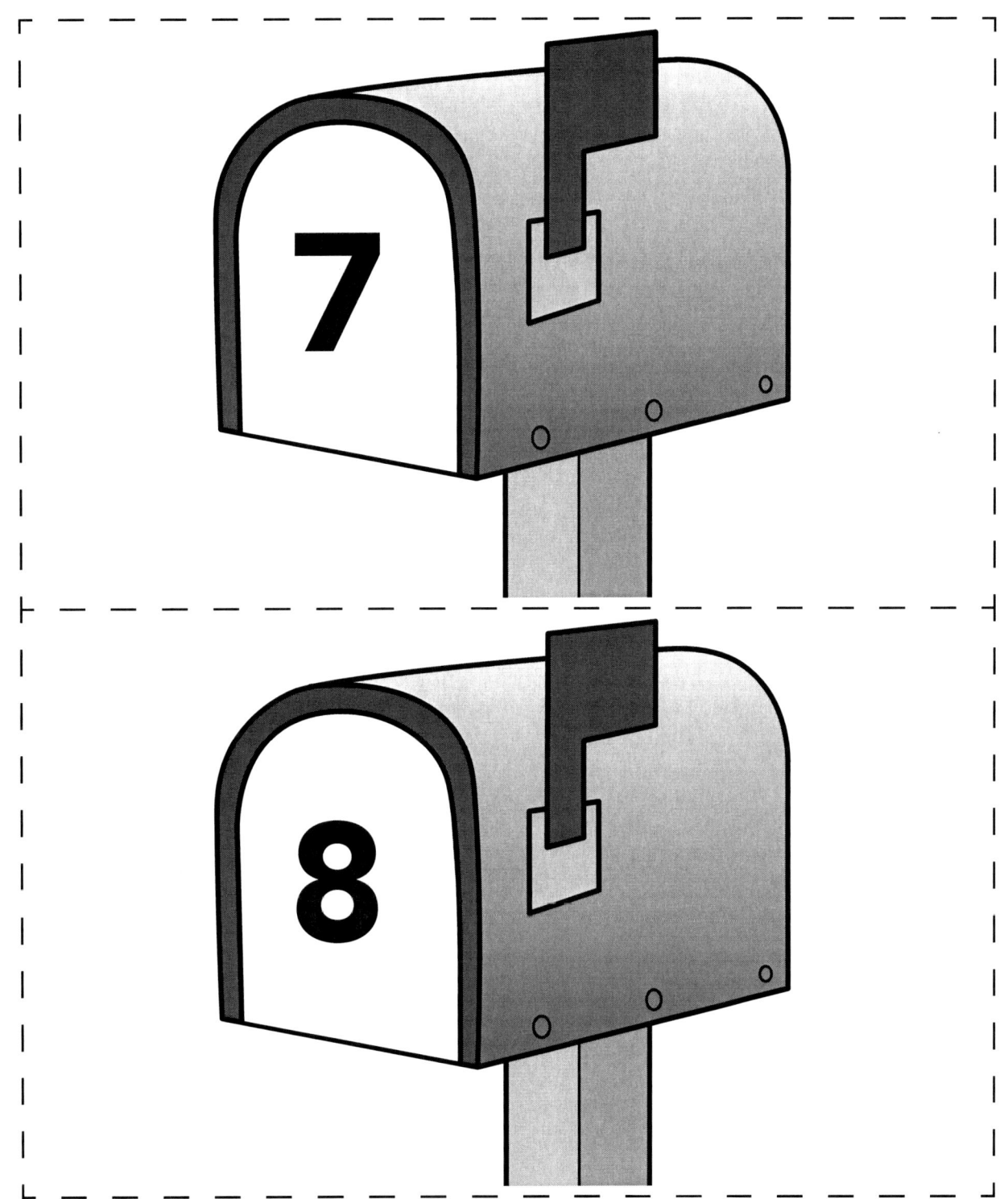

Operations and Algebraic Thinking

Mailbox
Templates (cont.)

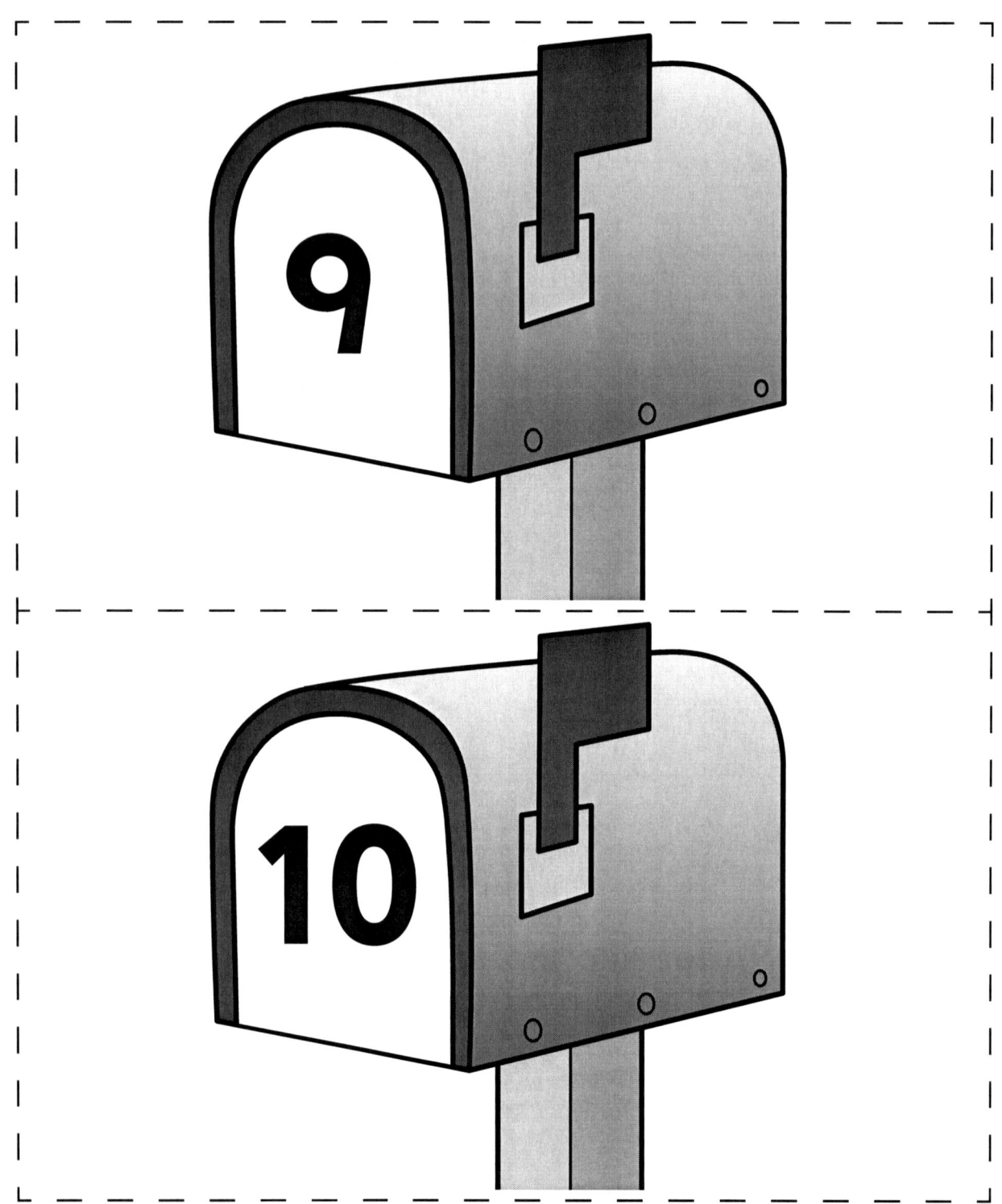

Operations and Algebraic Thinking

Mailbox
Templates (cont.)

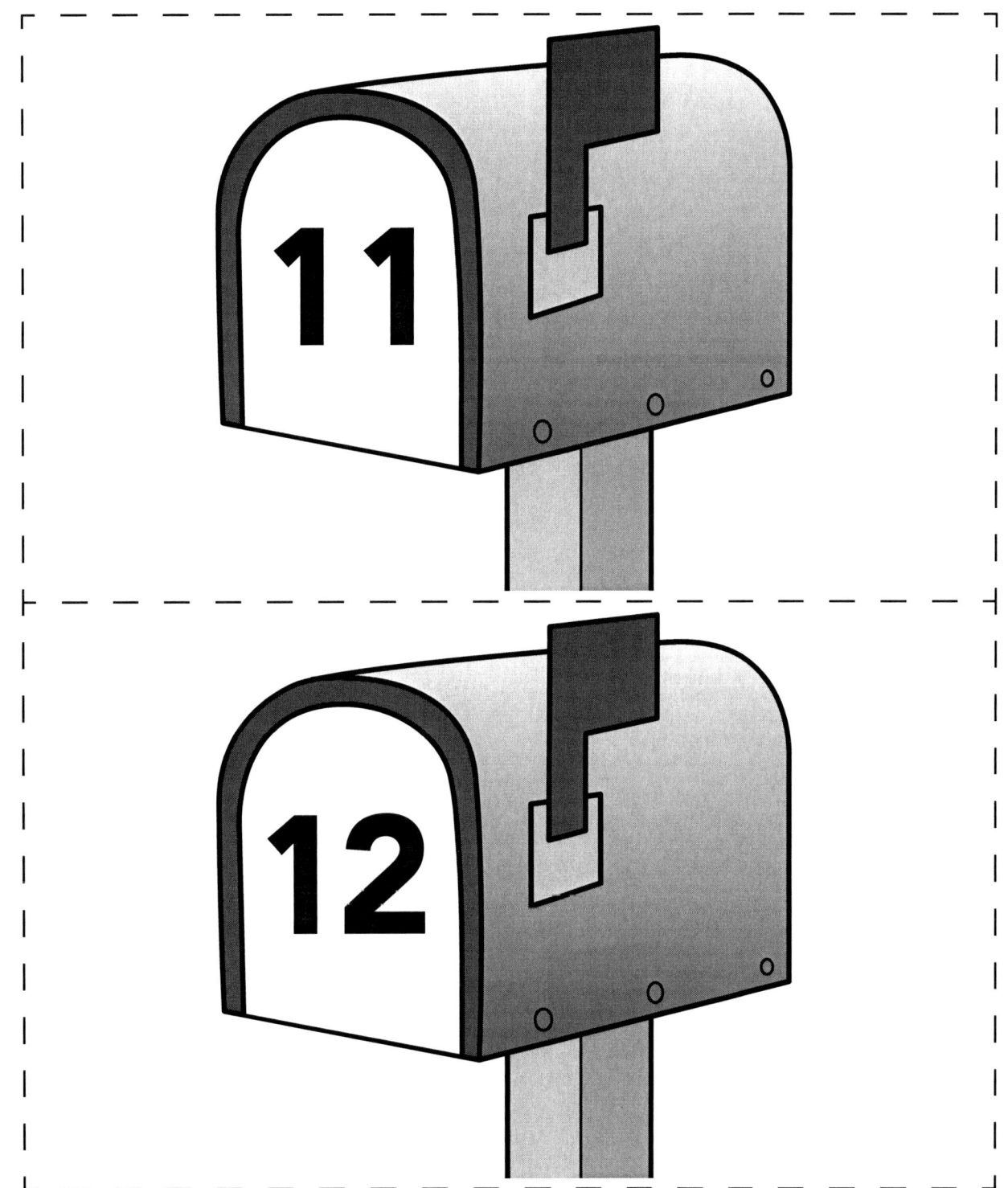

Mailbox
Templates (cont.)

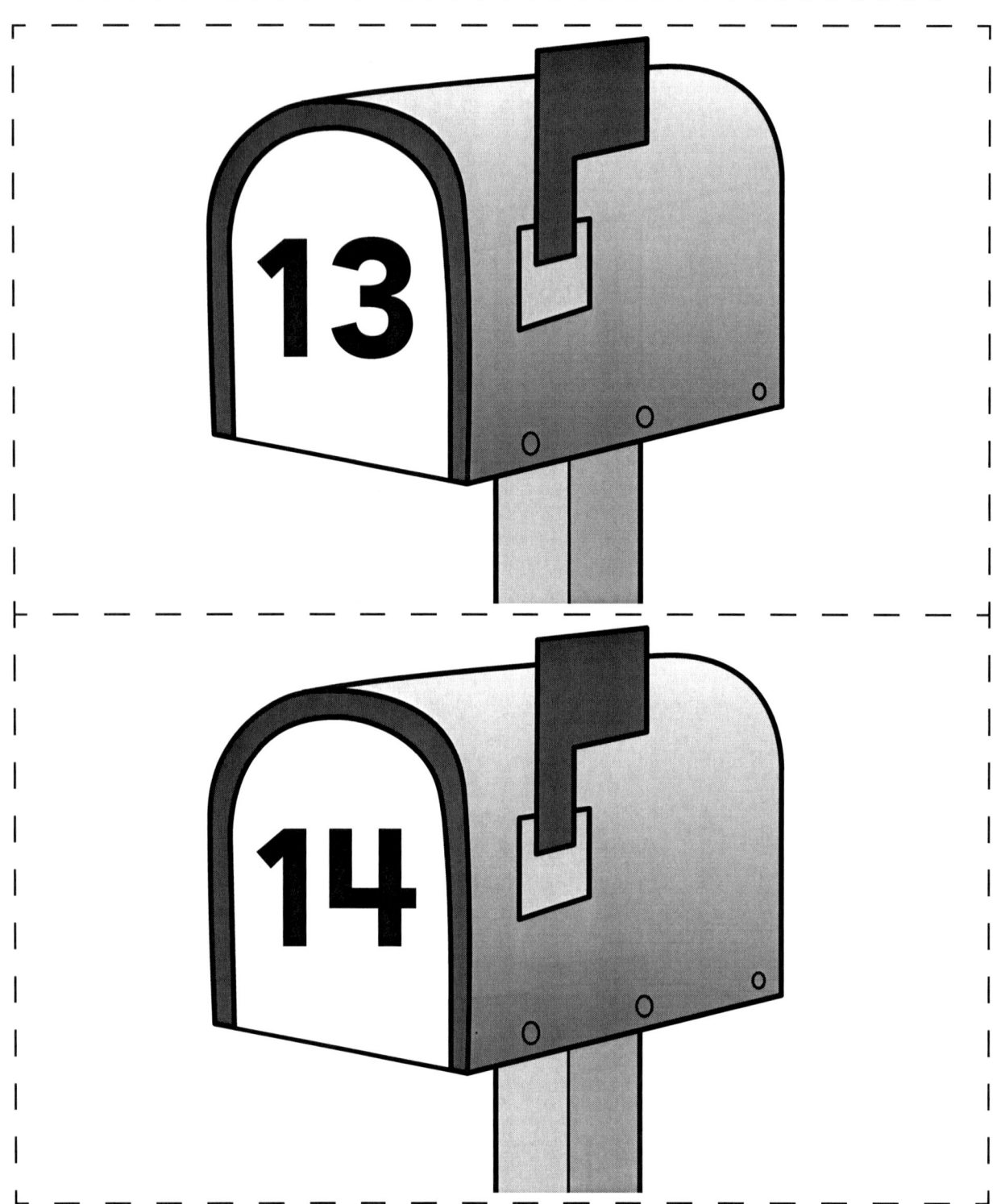

Operations and Algebraic Thinking

Mailbox
Templates (cont.)

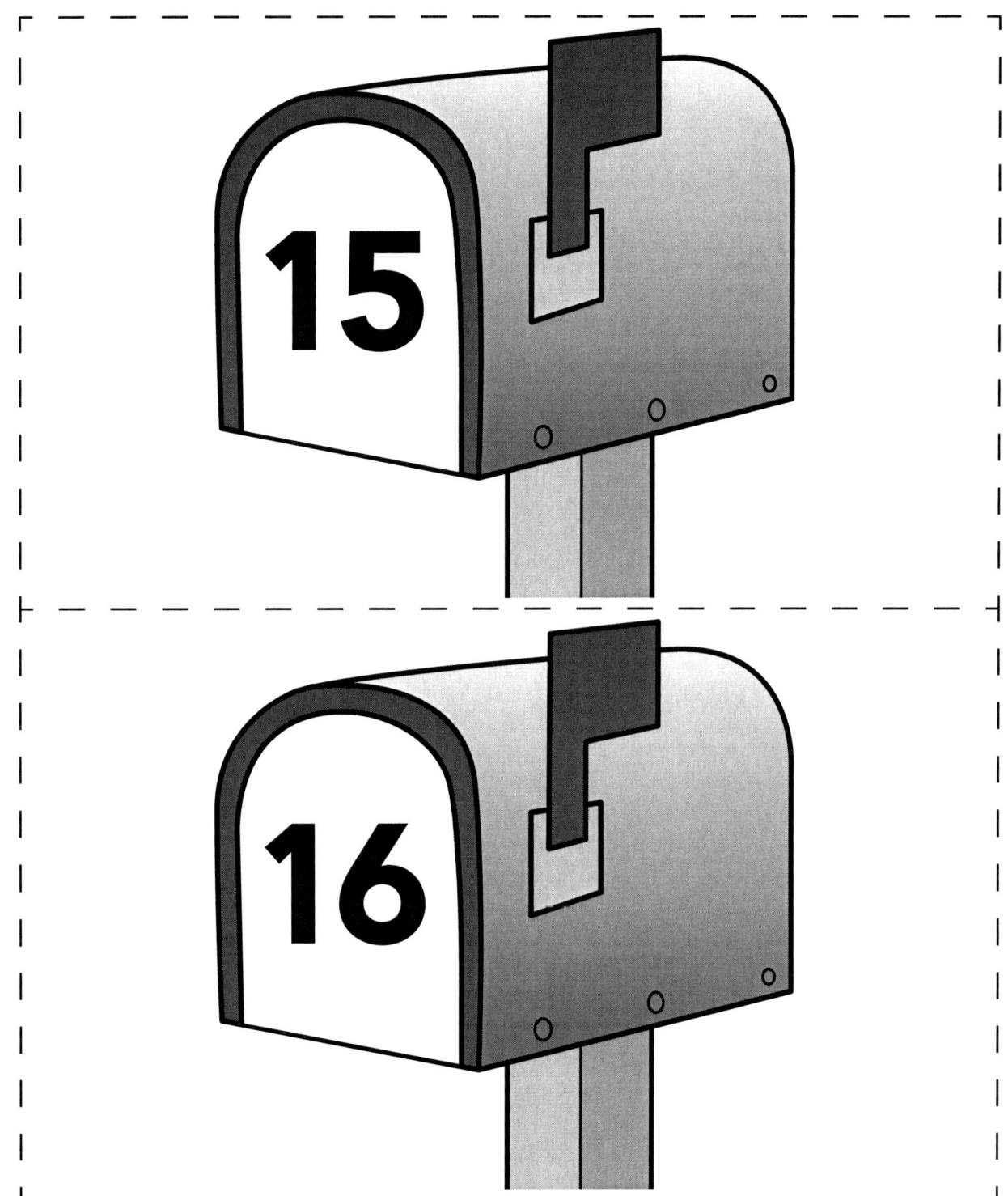

Operations and Algebraic Thinking

Mailbox
Templates (cont.)

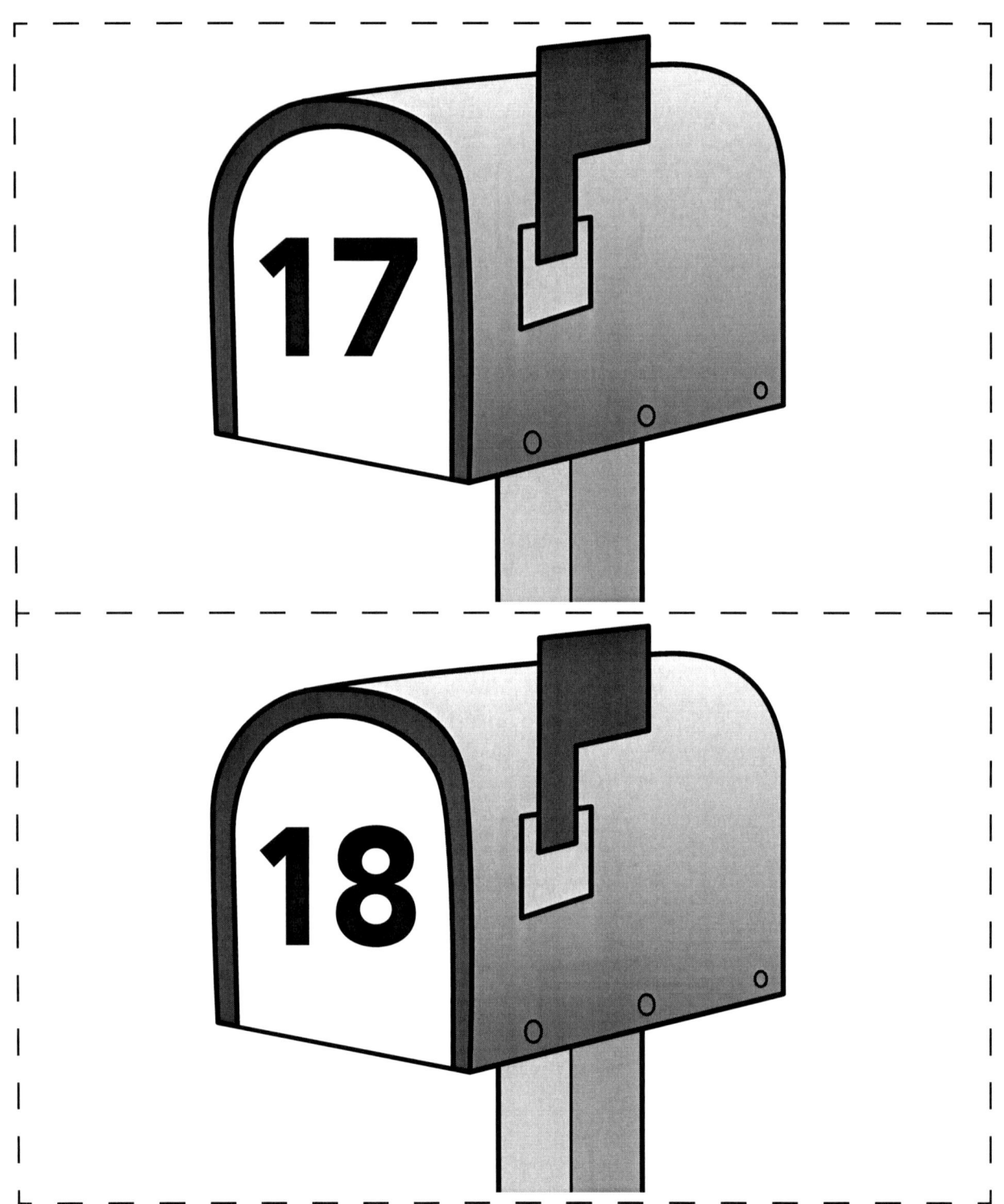

Operations and Algebraic Thinking

Mailbox
Templates (cont.)

© Shell Education #51288—Math Games: Skill-Based Practice 27

Operations and Algebraic Thinking

Envelope
Game Cards

Directions: Copy and cut out one set of game cards. Glue each card onto a white envelope.

I have 3 coins in my bank. Mom gave me 11 more coins. How many coins do I have now?

Mandy has 13 dollars. She gives 5 dollars to her little brother. How many dollars does she have now?

Envelope
Game Cards (cont.)

Bill made 3 drawings for his dad and 3 for his mom. How many drawings did he make in all?

Pat sees 7 kids playing on the playground. Then, 2 more kids come to play. How many kids are now playing on the playground?

Operations and Algebraic Thinking

Envelope
Game Cards (cont.)

I read 5 books last week and 2 more books this week. How many books did I read altogether?

Sam has 11 toy cars. For his birthday present, his mom bought him 1 more car. How many cars does Sam have?

Envelope
Game Cards (cont.)

Kari ate 2 brownies, and then she ate 2 more brownies. How many brownies did she eat?

Tami finds 7 leaves and 12 sticks on the playground. Altogether, how many leaves and sticks did she find?

Envelope
Game Cards (cont.)

Mom gave me 7 envelopes to stamp. Then she gave me 10 more envelopes to stamp. How many envelopes did I stamp in all?

I had 5 stuffed animals on my bed. I gave 4 stuffed animals to my little brother. How many stuffed animals do I have left?

Envelope
Game Cards (cont.)

Mom packed me 4 things for lunch: a sandwich, an apple, a cookie, and juice. I ate my sandwich and my apple. How many things do I have left?

I learned 9 new words last week and 2 more new words this week. How many new words did I learn altogether?

Operations and Algebraic Thinking

Envelope
Game Cards (cont.)

I am going to watch 2 movies this weekend and 1 movie next weekend. How many movies am I going to watch altogether?

I have 12 shirts in my closet. When I try them on, 7 shirts no longer fit me. How many shirts still fit me now?

Envelope
Game Cards (cont.)

I like to set the kitchen table for Mom. I put 2 plates on the table for Mom and Dad. I put 8 plates on the table for the kids. How many plates are there altogether?

Sally has 9 blue crayons on her desk and 6 red crayons. How many crayons does she have in all?

Operations and Algebraic Thinking

Envelope
Game Cards (cont.)

There are 8 boys and 8 girls in our class. How many students are there in all?

Mom says I can invite 10 girls and 8 boys to my party. How many boys and girls will I invite altogether?

Envelope
Game Cards (cont.)

Donnie has 7 red marbles and 6 blue marbles. How many marbles does he have altogether?

Operations and Algebraic Thinking

I Can Solve It!

Domain

Operations and Algebraic Thinking

Standard

Solve word problems that call for addition of three whole numbers whose sum is less than or equal to 20 (e.g., by using objects, drawings, and equations to represent the problem).

Number of Players

2 Players

Materials

- *I Can Solve It! Word Problem Cards* (pages 40–41)
- empty cans (e.g., soup cans, coffee cans, cylinder-shaped chip cans)
- large labels or a permanent marker
- drawing paper
- pencils
- number cubes
- manipulatives

GET PREPARED!

- Copy and cut out one set of *I Can Solve It! Word Problem Cards* for each pair of players.

- Prepare one empty can for each pair of players by writing *I can solve it!* on every can, using a permanent marker, or by placing a label containing the phrase *I can solve it!* onto the can. Check for any sharp edges taht may cuase injury to players.

- Collect one sheet of drawing paper and one pencil for each player.

- Collect one number cube and an assortment of manipulatives for each pair of players.

I Can Solve It! (cont.)

Game Directions

1. Distribute materials to players.

2. Each pair of players places their set of *I Can Solve It! Word Problem Cards* into their empty, labeled can and places the can between them. Players then set aside their pencils and sheets of drawing paper, and place the manipulatives they will be sharing in an easily accessible location for both players.

3. Players take turns rolling a number cube. The player who rolls the higher number is Player 1.

4. Player 1 draws a card from the can, reads it aloud, and then solves the word problem, using manipulatives or a drawing.

5. If Player 1 can *prove* the correct answer with a drawing or through the use of manipulatives, he or she keeps that card in his or her own "winning pile."

6. Player 2 takes a turn, repeating the process in steps 4 and 5.

7. The player with more cards in his or her winning pile at the end wins!

Operations and Algebraic Thinking

I Can Solve It!
Word Problem Cards

Directions: Copy and cut out one set of cards for each pair of players.

In my school bag, I have 3 pencils, 4 crayons, and 2 pens. How many writing tools do I have in my bag?	At the school library, I checked out 2 books about horses, 2 books about sports, and 1 book about camping. How many books did I check out altogether?
Mandy has 5 goldfish, 2 dogs, and 1 cat. How many animals does she have altogether?	Teddy has 3 blue shirts, 5 white shirts, and 3 green shirts. How many shirts does he have?

I Can Solve It!
Word Problem Cards (cont.)

To solve a problem, I used 6 yellow cubes, 7 orange cubes, and 3 brown cubes. How many cubes did I use altogether?

Kendra found 8 pennies, 8 dimes, and 1 nickel under the couch. How many coins did she find in all?

I reached into the candy bag and found 2 red candies, 6 green candies, and 8 yellow candies. How many candies did I find in all?

I helped decorate 7 white cupcakes, 4 green cupcakes, and 3 pink cupcakes. How many cupcakes did I decorate in all?

For our art project, I chose 9 pieces of white paper, 3 pieces of blue paper, and 2 pieces of purple paper. How many pieces of paper do I have in all?

For my new bracelet, I will use 6 white beads, 6 black beads, and 2 red beads. How many beads will I use in all?

Operations and Algebraic Thinking

Mix-Up, Match-Up

Domain X+Y=?
Operations and Algebraic Thinking

Standard
Apply properties of operations and strategies to add and subtract. Example: if 8 + 3 = 11 is known, then 3 + 8 = 11 is also known (commutative property).

Number of Players
2 Players

Materials
- *Mix-Up, Match-Up Game Board* (pages 44–45)
- *Mix-Up, Match-Up Game Cards* (page 46 and mixupcards2.pdf from Digital Resource CD)
- separate colors of game marker pieces (e.g., two-color counters, small colored cubes)
- number cubes

GET PREPARED!

- Copy and cut out a set of the *Mix-Up, Match-Up Game Cards* for each player.
- Copy and cut out the *Mix-Up, Match-Up Game Board* for each pair of players.
- Collect one number cube for each pair of players.
- Collect 15 game pieces for each player.

Game Directions

1. Distribute materials to players.

2. Players roll a number cube. The player who rolls the higher number is Player 1.

3. Players shuffle the *Mix-Up, Match-Up Game Cards* and place them facedown in a pile between each other.

4. Player 1 draws five cards from the deck. Player 2 also draws five cards from the deck.

42 #51288—*Math Games: Skill-Based Practice* © Shell Education

Mix-Up, Match-Up (cont.)

5 Player 1 searches through his or her *Mix-Up, Match-Up Game Cards*. Each card illustrates the commutative property for a number sentence on the *Mix-Up, Match-Up Game Board*. Player 1 strategically chooses one of his or her five cards to match it to a sentence on the board that illustrates the card's commutative property. When the sentence is found, the player puts one of his or her game pieces on the sentence. The goal is to make three in a row vertically, horizontally, or diagonally. Once a game piece is placed on the board, the used card goes into the "discard" pile.

6 Player 2 repeats steps 4 and 5.

7 Player 1 draws a card to have a total of five cards in his or her hand.

8 Either player can use his or her card to block the other player from scoring three in a row. If a player has the correct card, he or she places it in the correct space to prevent the other player from having three game markers in a row. .

9 Play continues until a player has placed 3 game pieces in a row.

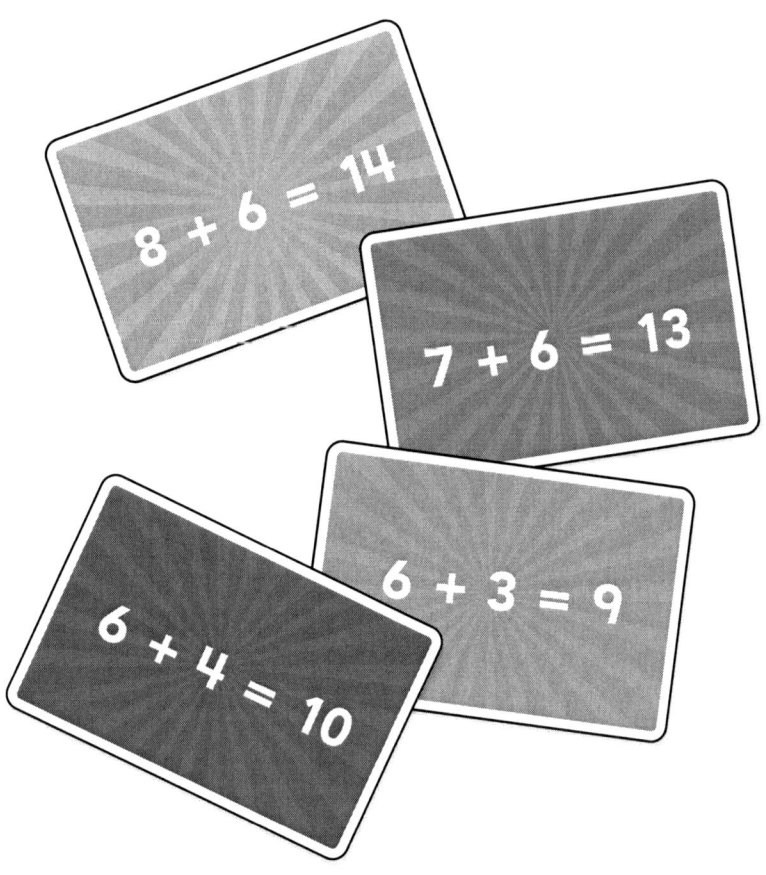

Operations and Algebraic Thinking

Mix-Up, Match-Up
Game Board

Directions: Copy and cut out one game board. Tape it to the game board on page 45.

Mix-Up,

3 + 6 = 9	8 + 7 = 15	6 + 2 = 8
6 + 7 = 13	6 + 8 = 14	7 + 6 = 13
3 + 7 = 10	1 + 9 = 10	6 + 4 = 10
8 + 4 = 12	4 + 3 = 7	4 + 8 = 12
2 + 9 = 11	6 + 3 = 9	7 + 5 = 12

tape here

Mix-Up, Match-Up
Game Board (cont.)

Match-Up

7 + 8 = 15	3 + 4 = 7	9 + 6 = 15
8 + 9 = 17	7 + 3 = 10	9 + 2 = 11
8 + 6 = 14	4 + 6 = 10	5 + 7 = 12
6 + 9 = 15	2 + 6 = 8	8 + 5 = 13
9 + 1 = 10	5 + 8 = 13	9 + 8 = 17

Operations and Algebraic Thinking

Mix-Up, Match-Up
Game Cards

Directions: Copy and cut out two sets of cards for each pair of players.

6 + 3 = 9	8 + 7 = 15
7 + 6 = 13	9 + 8 = 17
3 + 7 = 10	6 + 8 = 14
4 + 8 = 12	6 + 9 = 15
2 + 9 = 11	1 + 9 = 10
7 + 5 = 12	6 + 2 = 8

It's a Fact!

Domain
Operations and Algebraic Thinking

Standard
Understand subtraction as an unknown-addend problem. For example, subtract 10 − 8 by finding the number that makes 10 when added to 8.

Number of Players
2 Players

Materials
• *It's a Fact! Game Board* (page 49) • *It's a Fact! Game Cards* (page 50) • number cubes

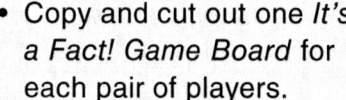

- Copy and cut out one *It's a Fact! Game Board* for each pair of players.
- Copy and cut out the *It's a Fact! Game Cards* for each pair of players.
- Collect one number cube for each pair of players.

Game Directions

1 Distribute materials to players.

2 Players take turns rolling a number cube. The player who rolls the lower number is Player 1.

3 Players shuffle the *It's a Fact! Game Cards* and place them facedown between them. The *It's a Fact! Game Board* should also sit between both players. Each player receives one "remove" card to strategically use if an equation can't be formed and a played number card needs to be removed from the board. The card may only be used once per player during each game. To use the card, simply place it near the deck and remove one number card from the board, placing it at the bottom of the deck and replace it with a new number card.

It's a Fact! (cont.)

4. Player 1 draws a card and places it faceup in any open spot on the game board to start forming a subtraction problem.

5. Player 2 draws a card and also places it faceup in any open spot with the goal of making a subtraction problem. Players do not have assigned spots on the game board and are free to place their cards anywhere—even on an equation already started by the opposing player.

6. Once the board starts to get filled with cards, if a player cannot make a play on his or her turn because a subtraction problem cannot be formed, that "unplayable" card gets placed at the bottom of the card deck. The player then draws a new card from the top of the deck and attempts to form a problem with that card. Players repeat this process until an equation is made on his or her turn while alternating turns with the other player.

7. The first player to form a correct subtraction equation takes all three of the game cards that compose the subtraction problem and keeps them in his or her own "winning pile."

8. Once a series of cards are removed from the game board and placed in players' winning piles, the three open spaces may be used again to create another subtraction problem.

9. When the teacher calls, "Time," the player with more cards in his or her winning pile wins!

Operations and Algebraic Thinking

It's a Fact!
Game Board

Directions: Copy and cut out the game board for each pair of students.

It's a Fact

☐ − ☐ = ☐

☐ − ☐ = ☐

☐ − ☐ = ☐

☐ − ☐ = ☐

☐ − ☐ = ☐

☐ − ☐ = ☐

☐ − ☐ = ☐

Operations and Algebraic Thinking

It's a Fact!
Game Cards

Directions: Copy and cut out the cards for each pair of players.

1	2		1	2
3	4		3	4
5	6		5	6
7	8		7	8
9	10		9	10
1	2		1	2
3	4		3	4
5	6		5	6
7	8		7	8
9	10		9	10

Remove a Card	Remove a Card

Speedy Counters

Domain

Operations and Algebraic Thinking

Standard

Relate counting to addition and subtraction (e.g., by counting on 2 to add 2).

Number of Players

4 Players

Materials

- *Speedy Counters Race Track* (pages 52–53)
- *Speedy Counters Race Cards* (pages 54–59)
- game piece for each player (e.g., two-color counters)
- number cubes

GET PREPARED

- Copy and assemble one *Speedy Counters Race Track* for each group of 4 players.
- Copy and cut out one set of *Speedy Counters Race Cards* for each group of 4 players.
- Collect one game piece for each player.
- Collect one number cube for each group of 4 players.

Game Directions

1. Each player rolls the number cube to determine who the "race caller" will be. The race caller should change each round and will be determined by the player who rolls the highest number at the start of every round. The race caller's role includes shuffling the *Speedy Counters Race Cards* and placing them facedown in front of all three players. Throughout the game, he or she also calls out each race card. The race caller must speak clearly and listen closely to make sure the racers are giving accurate answers.

2. Players place their game pieces at their designated "Start" points on the racetrack.

3. The race caller shuffles the cards and places them facedown in front of all 3 players. He or she draws a card from the top and reads the math problem (e.g., 9 + _____ = 11) using a clear voice that can be heard by all of his or her teammates.

4. The first player to state the correct response moves his or her game piece forward one space on the racetrack. All players must state the correct response before the race caller moves on to the next card, but only the first racer to state the fact moves forward.

5. The first player to get to the "Finish" space wins!

Operations and Algebraic Thinking

Speedy Counters
Race Track

Directions: Cut out the game board. Tape it to the other game board on page 53.

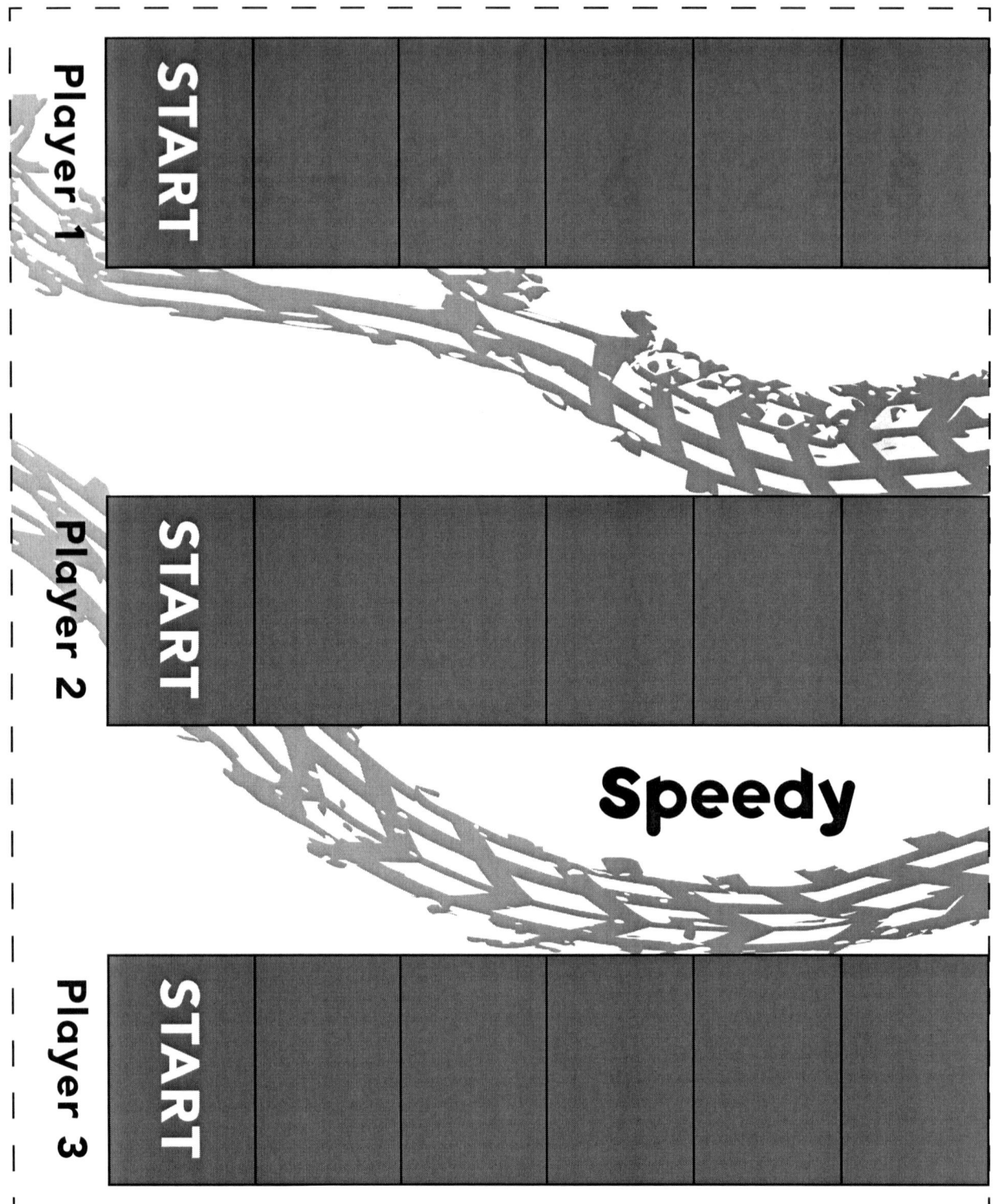

Operations and Algebraic Thinking

Speedy Counters
Race Track (cont.)

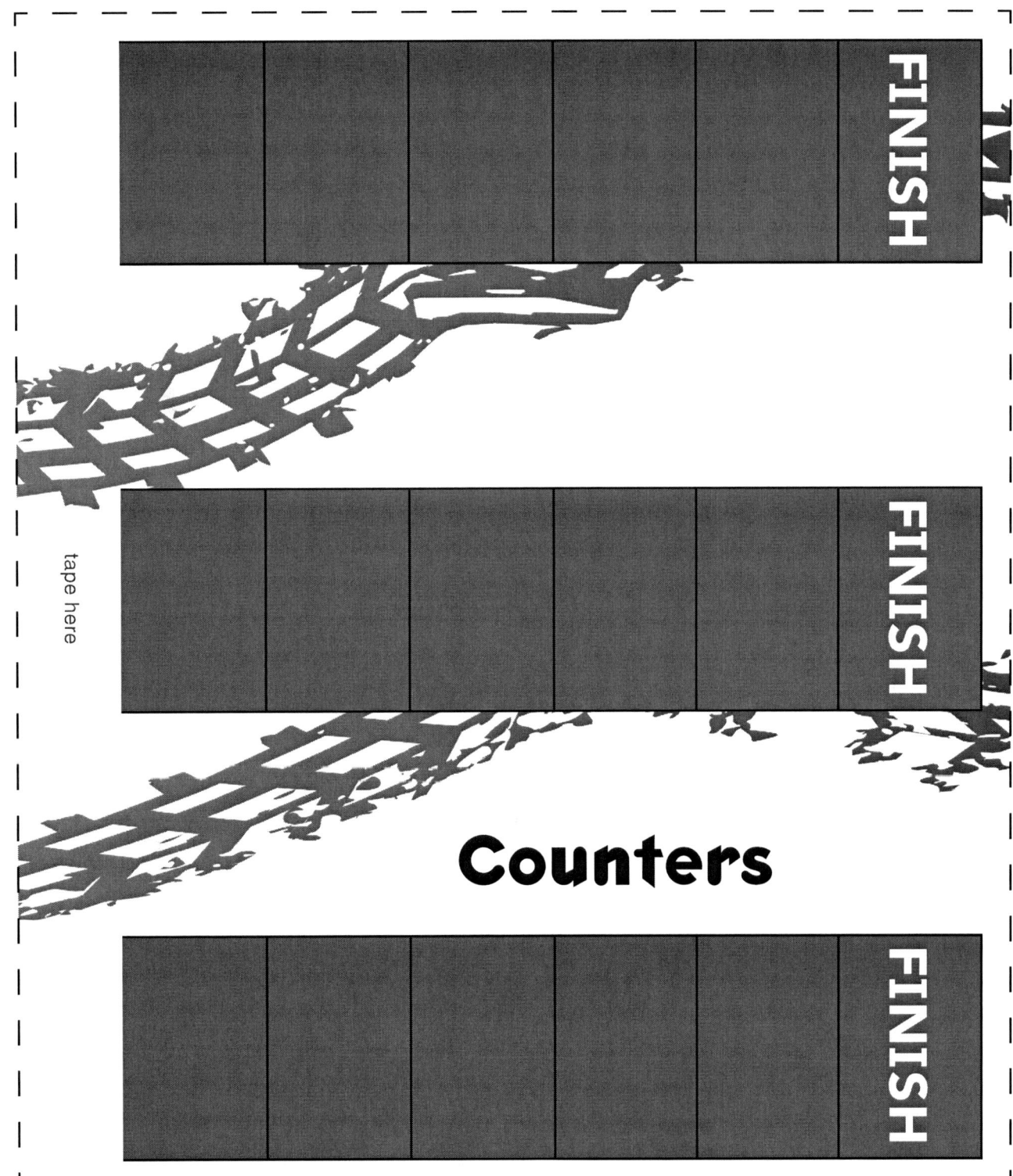

Operations and Algebraic Thinking

Speedy Counters
Race Cards

Directions: Copy and cut out one set of cards for each group of players.

9 + ___ = 11

___ + 7 = 10

8 + ___ = 9

___ + 4 = 7

5 + ___ = 8

4 + ___ = 8

Speedy Counters
Race Cards (cont.)

9 − ___ = 6

10 − ___ = 8

___ + 13 = 15

16 − ___ = 13

12 − ___ = 9

11 − ___ = 9

Operations and Algebraic Thinking

Speedy Counters
Race Cards (cont.)

5 + ___ = 6

___ + 17 = 18

8 + ___ = 11

___ + 3 = 7

6 + ___ = 8

4 + ___ = 5

Speedy Counters
Race Cards (cont.)

9 − ___ = 5

10 − ___ = 9

___ + 12 = 15

17 − ___ = 15

12 − ___ = 8

13 − ___ = 12

Operations and Algebraic Thinking

Speedy Counters
Race Cards (cont.)

9 + ___ = 13

___ + 7 = 11

8 + ___ = 12

___ + 2 = 7

7 + ___ = 8

15 + ___ = 18

Operations and Algebraic Thinking

Speedy Counters
Race Cards (cont.)

11 − ___ = 8

10 − ___ = 6

___ + 13 = 16

17 − ___ = 12

12 − ___ = 8

10 − ___ = 9

Operations and Algebraic Thinking

Climb to the Top

Domain
Operations and Algebraic Thinking

Standard
Add and subtract within 20, demonstrating fluency for addition and subtraction within 10. Use strategies such as counting on, make a ten, decomposing a number leading to ten, using the relationship between addition and subtraction, and creating equivalents to easier or known sums.

Number of Players
2 Players

Materials
- *Climb to the Top Spinner* (page 61)
- *Climb to the Top Game Board* (pages 62–63)
- pencils and paperclips
- game piece for each player (e.g., two-color counters, small colored cubes)
- number cubes

GET PREPARED!

- Copy and cut out a *Climb to the Top Spinner* and a *Climb to the Top Game Board* for each pair of players.
- Collect a game piece for each player and one paperclip, pencil, and number cube for each pair of players.

Game Directions

1. Players take turns rolling a number cube. The player who rolls the higher number is Player 1.

2. Player 1 flicks the paperclip on the spinner and moves the designated amount of spaces on the game board to solve the math problem written on the space.

3. If correct, the player keeps his or her game piece on that space and explains the strategy used. For example, if the player lands on "7 + 8," he or she must explain the thinking used, for example, "I know that 7 + 7 is a double that adds to 14, plus one more makes 15. I used the strategy doubles plus one." If incorrect, the player moves back one space.

4. Player 2 repeats steps 2 and 3.

5. The first player to reach *Finish* wins!

Operations and Algebraic Thinking

Climb to the Top
Spinner

Directions: Copy and cut out the spinner for each pair of players. For steps on how to assemble this spinner, see page 10.

Operations and Algebraic Thinking

Climb to the Top
Game Board

Directions: Cut out the game board. Tape it to the game board on page 63.

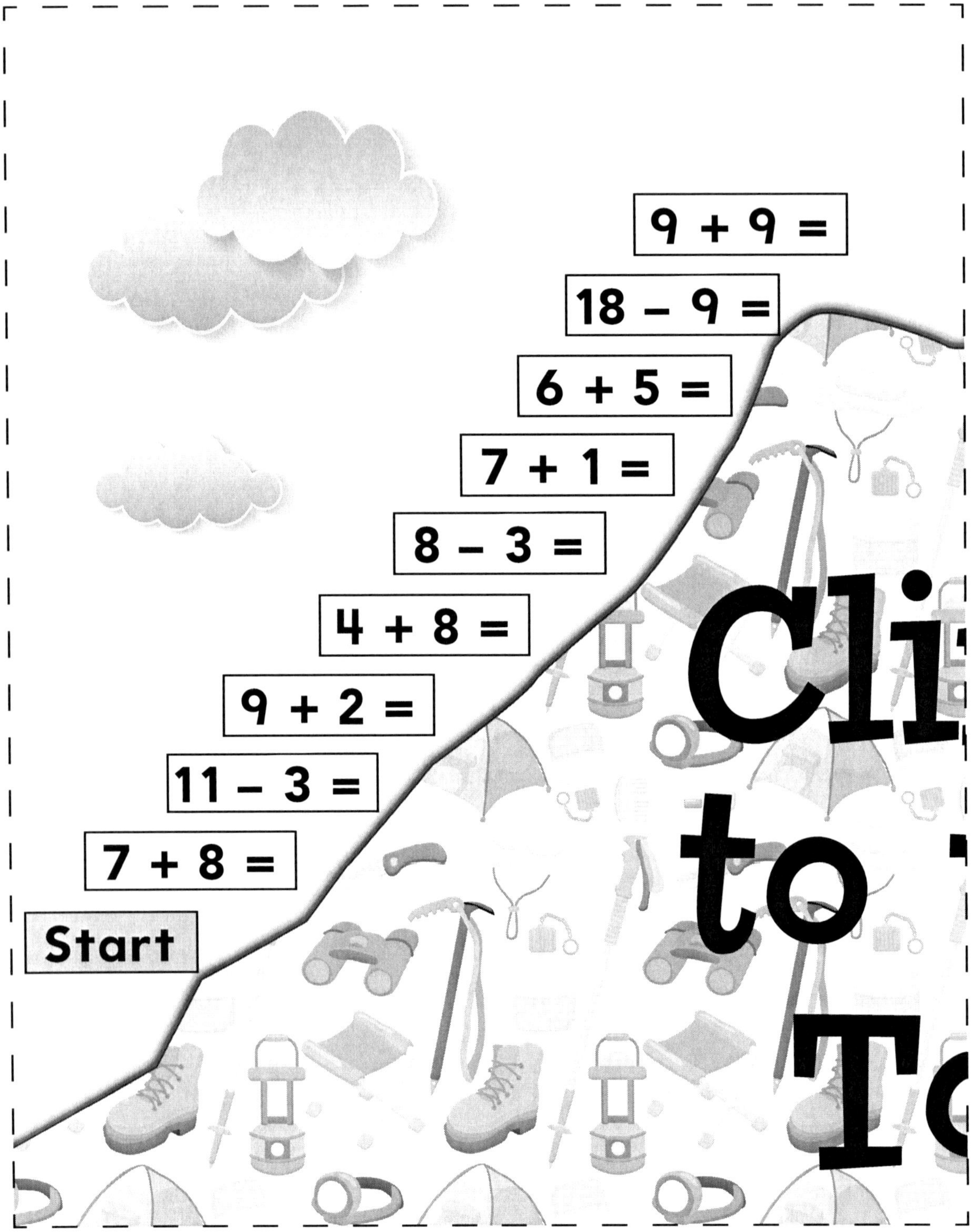

9 + 9 =

18 − 9 =

6 + 5 =

7 + 1 =

8 − 3 =

4 + 8 =

9 + 2 =

11 − 3 =

7 + 8 =

Start

Climb to the Top
Game Board (cont.)

5 + 4 =
6 − 4 =
8 − 5 =
11 + 4 =
7 + 7 =
9 − 2 =
13 − 6 =
3 + 7 =
12 + 5 =
Finish

Operations and Algebraic Thinking

True or False

Domain

Operations and Algebraic Thinking

Standard

Understand the meaning of the equal sign and determine if equations involving addition and subtraction are true or false. For example, which of the following equations are true and which are false? 6 = 6, 7 = 8 − 1, 5 + 2 = 2 + 5, 4 + 1 = 5 + 2.

Number of Players

4 Players

Materials

- *True or False Cards* (pages 65–66)
- number cubes

GET PREPARED!

- Copy and cut out one set of *True or False Cards* for each group of 4 players.
- Collect a number cube for each group of players.

Game Directions

1. Distribute materials to players. Players roll the number cube to determine who plays first. The player with the lowest roll is Player 1.

2. Player 1 shuffles the *True or False Cards* and distributes four cards facedown to each player.

3. Player 1 picks up the first card from his or her stack, determines if the statement is true or false, and explains why. If Player 1 is correct, he or she may keep the card. For example, if the player states that 4 + 6 = 8 + 2 is true and explains that both sides add up to 10, the player may keep the card in his or her "winning pile." If incorrect, the player must surrender the card to the discard pile.

4. The other players take turns in the same manner.

5. The player with the most cards at the end of the game wins.

True or False
Cards

Directions: Copy and cut out one set of cards for each group of players.

 6 + 2 = 10 − 2

 3 + 3 = 6 − 2

 9 − 2 = 4 + 3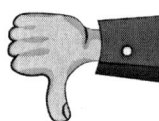

 15 − 5 = 10 + 0

 17 − 9 = 6 + 2

 17 − 5 = 8 + 4

12 + 3 = 15 − 3

16 − 2 = 13 + 3

True or False
Cards (cont.)

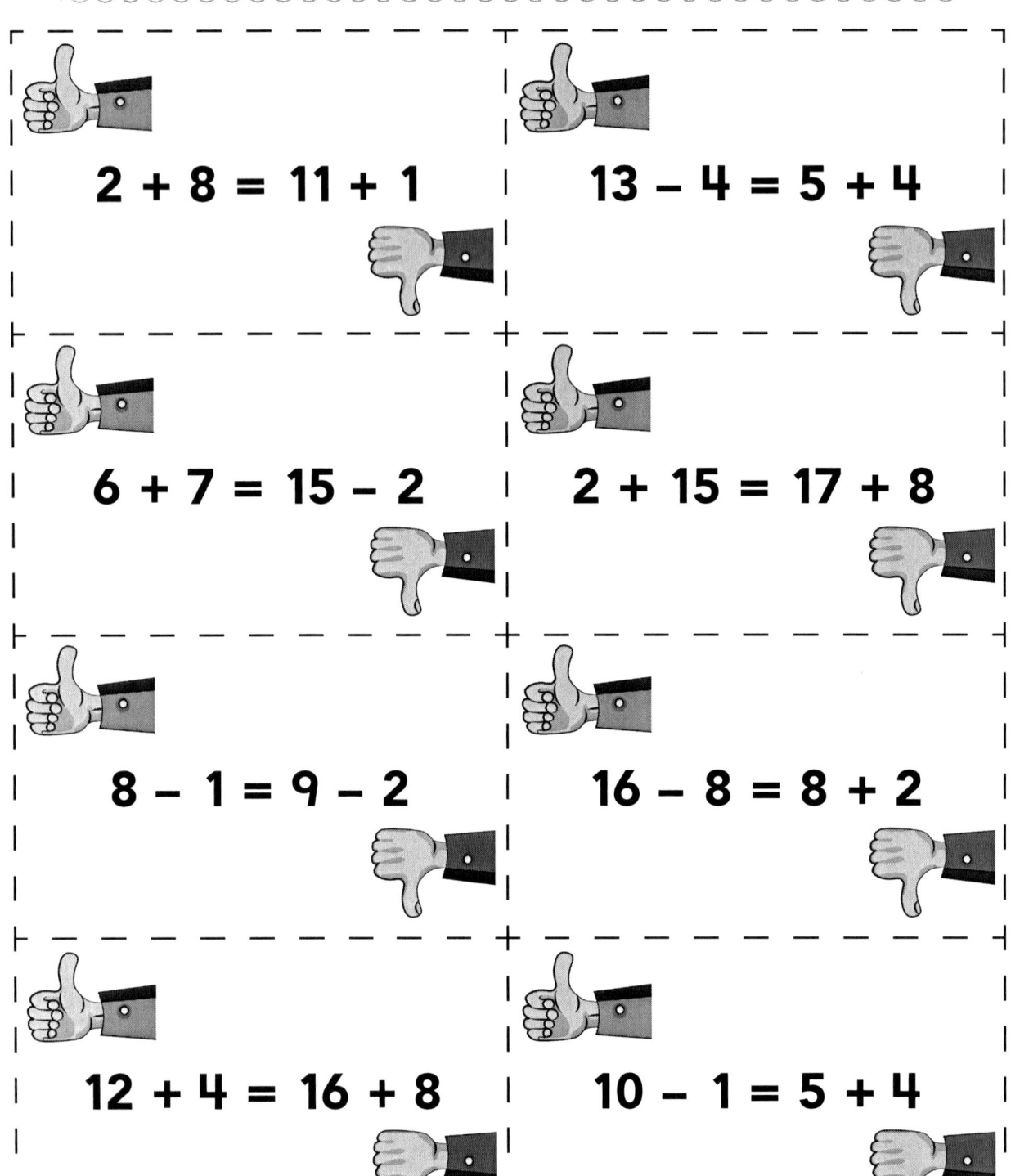

2 + 8 = 11 + 1

13 − 4 = 5 + 4

6 + 7 = 15 − 2

2 + 15 = 17 + 8

8 − 1 = 9 − 2

16 − 8 = 8 + 2

12 + 4 = 16 + 8

10 − 1 = 5 + 4

Unknown Number Game

Domain

Operations and Algebraic Thinking

Standard

Determine the unknown whole number in an addition or subtraction equation relating three whole numbers. For example, determine the unknown number that makes the equation true in each of the equations: 8 + ? = 11; 6 + 6 = ?; 5 = ? – 3.

Number of Players

2 Players

Materials

- *Missing Number Cards* (page 68)
- *Unknown Number Equation Cards* (pages 69–71)

GET PREPARED!

Copy and cut out a set of *Missing Number Cards* and a set of *Unknown Number Equation Cards* for each pair of players.

Game Directions

1. Player 1 shuffles the *Unknown Number Equation Cards* and places the cards facedown in a stack between the partners. Players arrange the *Missing Number Cards* faceup between them.

2. Player 1 draws an equation card from the top of the stack and places it face up. The player selects a *Missing Number Card* to make the equation true. For example, for 8 + ? = 11, the player would select the 3 card.

3. If Player 1 is correct, he or she keeps the equation card in a "winning pile." If Player 1 is not correct, Player 2 can try to choose a card to make the equation correct. If he or she succeeds, Player 2 keeps the equation card in his or her winning pile. If the equation is still not correct, it is placed at the bottom of the pile.

4. Players return the *Missing Number Cards* to their places after each round.

5. Player 2 repeats steps 2 and 3.

6. Play continues until all of the equation cards have been used. The player with more *Unknown Number Equation Cards* at the end of the game wins!

Operations and Algebraic Thinking

Missing Number
Cards

Directions: Copy and cut out one set of cards for each pair of students.

1	2	3	4
5	6	7	8
9	10	11	12
13	14	15	16
17	18	19	20

Operations and Algebraic Thinking

Unknown Number
Equation Cards

Directions: Copy and cut out one set of cards for each pair of players.

☐ + 7 = 12   15 − ☐ = 7

19 − ☐ = 11   ☐ + 5 = 20

☐ − 8 = 12   15 − ☐ = 6

© Shell Education #51288—Math Games: Skill-Based Practice

Operations and Algebraic Thinking

Unknown Number
Equation Cards (cont.)

8 − ▢ = 7   ▢ + 4 = 15

▢ + 12 = 20   ▢ − 12 = 5

3 + ▢ = 8   8 + ▢ = 15

Unknown Number
Equation Cards (cont.)

☐ + 3 = 17   ☐ − 7 = 5

19 − ☐ = 13   9 + ☐ = 14

14 − ☐ = 6   6 + ☐ = 18

The Next 3...

Domain
Numbers and Operations in Base Ten

Standard
Count to 120 starting at any number less than 120. In this range, read and write numerals and represent a number of objects with a written numeral.

Number of Players
Whole class divided into two teams

Materials
- index cards
- individual whiteboards

GET PREPARED!
Prepare a set of index cards (one card for each student) with a different random number on each. All numbers should be less than 117.

Collect individual whiteboards for each player.

Game Directions

1. Distribute materials to players.

2. A player from Team 1 goes to the front of the room and reads his or her number out loud to his or her team. For example, the player reads the number 39. Team 1 players at their desks write the next three numbers (40, 41, and 42) and then show their whiteboards to the other team.

3. While teammates are writing the next three numbers at their desks, the player at the front also writes the next three numbers on his or her own board.

4. Team 1 receives a tally mark for every correct player response on the team.

5. Team 2 takes a turn, repeating steps 2, 3, and 4.

6. The teacher keeps score on the board. The team with more tally marks at the end of 4 rounds wins.

Place Value

Domain

Numbers and Operations in Base Ten

Standard

Understand that the two digits of a 2-digit number represent amounts of tens and ones.

Number of Players

2 Players

Materials

- *Place Value Spinner* (page 74)
- *Place Value Game Board* (pages 75–76)
- number cubes
- paperclips and pencils
- game marker for each player (e.g., two-color counters, small colored cubes, mini erasers)

GET PREPARED

- Copy and cut out one *Place Value Game Board* and one *Place Value Spinner* for each pair of players.
- Collect one paperclip, one pencil, one number cube, and two game pieces for each pair of players.

Game Directions

1. Distribute materials to players.

2. Players take turns rolling the number cube. The player who rolls the lower number is Player 1.

3. Player 1 flicks the paperclip around the pencil in the center of the spinner and moves the indicated number of spaces.

4. Player 1 answers the question in the space occupied by his or her game marker. If the correct answer is given, the player stays on the space, but if incorrect, the player moves back one space.

5. Player 2 repeats steps 3 and 4.

6. The first player to reach the *End* space wins!

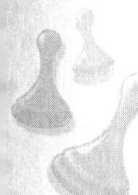

Place Value
Spinner

Directions: Copy and cut out one spinner for each pair of players. For steps on how to assemble this spinner, see page 10.

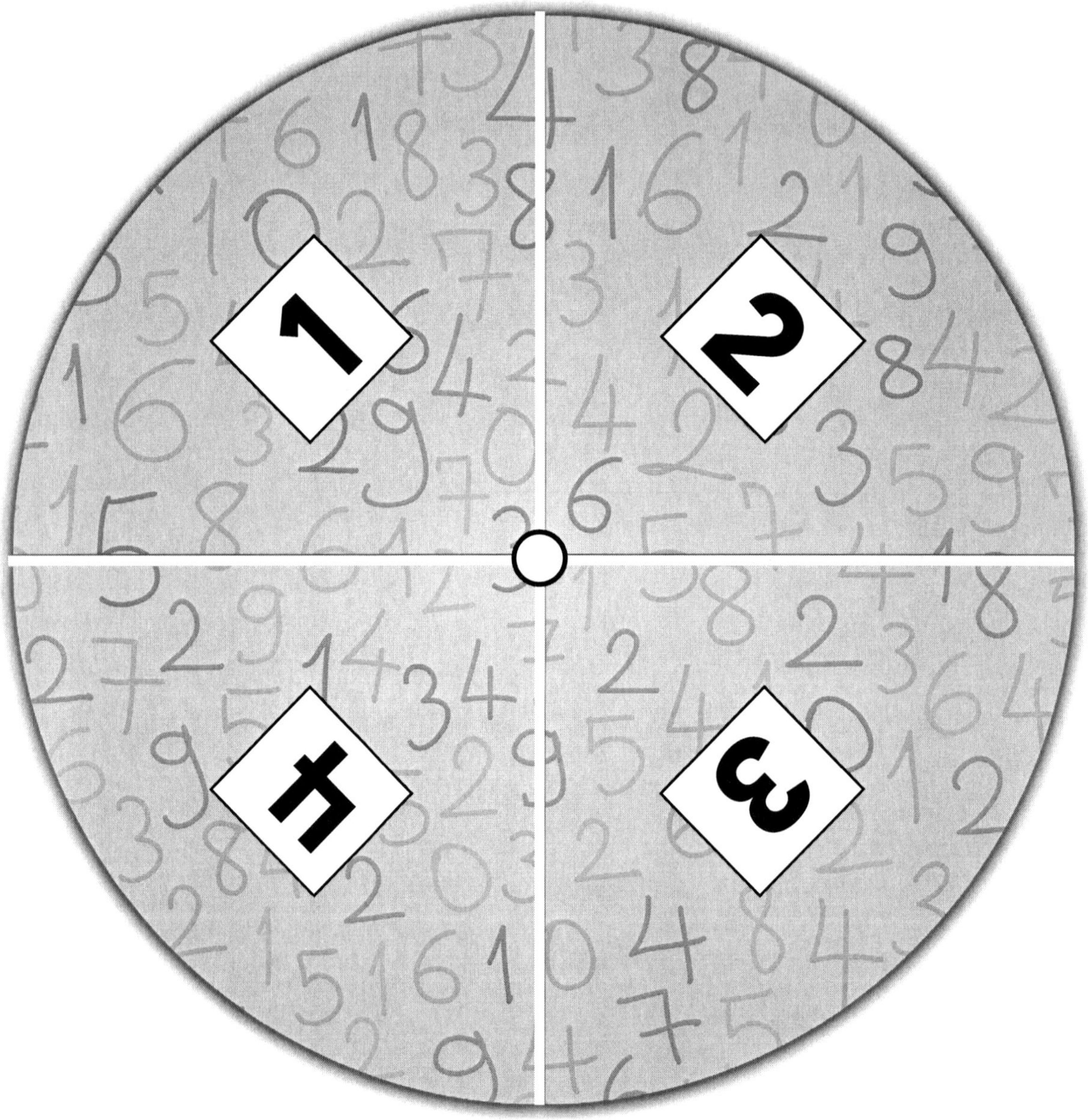

Place Value
Game Board

Directions: Copy and cut out the game board. Tape it to the game board on page 76.

START	How many tens? 29 = ? tens	What is my number? 3 tens and 2 ones
What is my number? 6 tens and 3 ones	What is my number? 8 tens and 9 ones	Move 1 space ahead
Move 1 space ahead	What is my number? 2 tens and 4 ones	What is my number? 3 tens and 12 ones
END	What is my number? 5 tens and 9 ones	What is my number? 6 tens and 14 ones

Place Value
Game Board (cont.)

Value

- How many tens?

  37 = ? tens

- How many ones?

  45 = ? ones

- How many ones?

  59 = ? ones

- What is my number?

  9 tens and 9 ones

- How many ones?

  47 = ? ones

- What is my number?

  2 tens and 8 ones

- What is my number?

  3 tens and 2 ones

- Move 1 space ahead

tape here

Roll It! Place Value Game

Domain

Numbers and Operations in Base Ten

Standard

Compare two 2-digit numbers based on the meanings of the tens and ones digits, recording the results of comparisons with the symbols >, =, and <.

Number of Players

2 Players

Materials

- *Roll It! Spinner* (page 78)
- *Roll It! Recording Page* (page 79)
- paperclips and pencils

GET PREPARED!

- Make one copy of the *Roll It! Spinner* and two copies of the *Roll It! Recording Page* for each pair of players.
- Collect one paperclip and one pencil for each pair of players.

Game Directions

1. Distribute materials to players.

2. Players each take one turn to flick the paperclip around the pencil in the center of the spinner. The player who spins the larger number is Player 1.

3. Player 1 flicks the spinner two times and records the numbers in the first row of each section of the *Roll It! Recording Page*, for example, 2 and 5.

4. Player 1 combines these numbers to create a 2-digit number in the second row of each section, for example, 25 or 52.

5. Player 2 spins two times, records the numbers in the same row as Player 1's number, and combines the two numbers to create a 2-digit number. For example, Player 2 spins 3 and 7, and then combines them to create 37 or 73.

6. Player 1 places a greater than, less than, or equal sign in the center column to compare the 2-digit numbers. If correct, Player 1 scores a point by placing a check in the "Point" column. If incorrect, no point is scored.

7. Players reverse roles for the next round.

8. The game ends when the *Recording Pages* are full. The player with more points wins!

Numbers and Operations in Base Ten

Roll It!
Spinner

Directions: Copy and cut out the spinner for each pair of players. For steps on how to assemble this spinner, see page 10.

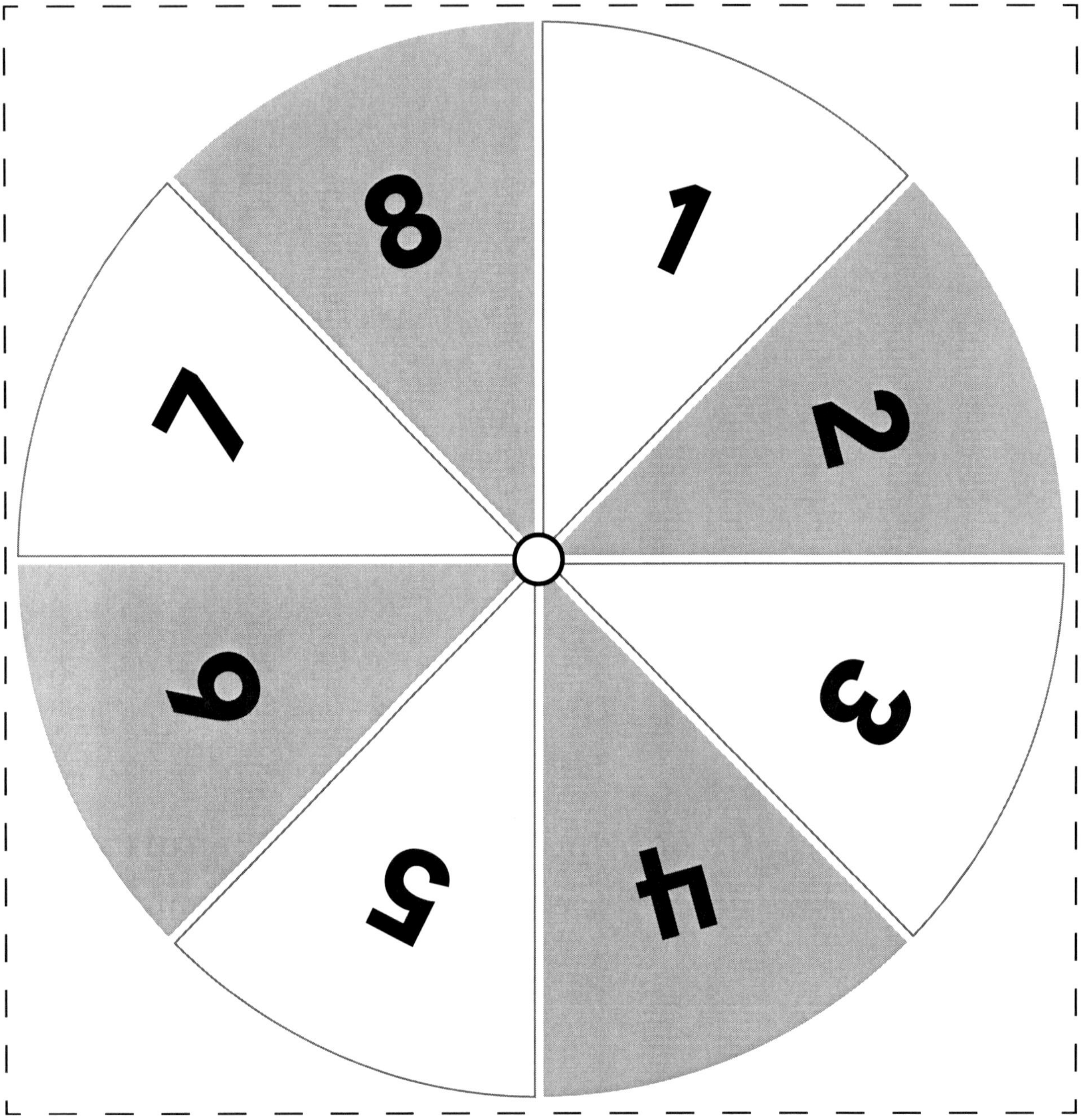

Name: _____   Date: _____

Roll It!
Recording Page

Directions: Use this chart to record your spins. The first section has been filled in for you as an example.

Rolls	Player 1		<, >, or =	Player 2		Points
	Spin 1	Spin 2		Spin 1	Spin 2	
Rolls	2	5		7	3	
2-digit	25		<	73		
Rolls						
2-digit						
Rolls						
2-digit						

Handy Math

Domain

Numbers and Operations in Base Ten

Standard

Add within 100, including adding a 2-digit number and a 1-digit number and adding a 2-digit number and a multiple of 10, using concrete models or drawings and strategies based on place value, properties of operations, and/or the relationship between addition and subtraction; relate the strategy to a written method and explain the reasoning used. Understand that in adding 2-digit numbers, one adds tens and ones; and sometimes it is necessary to compose a ten.

Number of Players

Whole class divided into two teams

Materials

- *Handy Math Equation Cards* (pages 81–86)
- base ten blocks or concrete models
- individual whiteboards

GET PREPARED!

- Copy and cut out a set of *Handy Math Equation Cards*.
- Collect base ten blocks or concrete models and whiteboards for each player.

Game Directions

1. Distribute materials to players and divide the *Handy Math Equation Cards* into 2 piles.

2. Pull the top card and hold it up for the class to see. Players work together in their teams to solve the problem and record it on their whiteboards. Players may use base ten blocks or whiteboards to create a drawing of the problem.

3. After a minute or so, the leader selects a player from Team 1 to show and explain how the team solved the problem. If correct, Team 1 receives a point. If incorrect, the leader calls on someone from Team 2 to provide a correct answer and explain how the problem was solved. If correct, Team 2 receives the point. The teacher keeps score on the board.

4. The game continues until 8 rounds have been completed. The team with more points wins!

Numbers and Operations in Base Ten

Handy Math
Equation Cards

Directions: Copy and cut out one set of cards for the class.

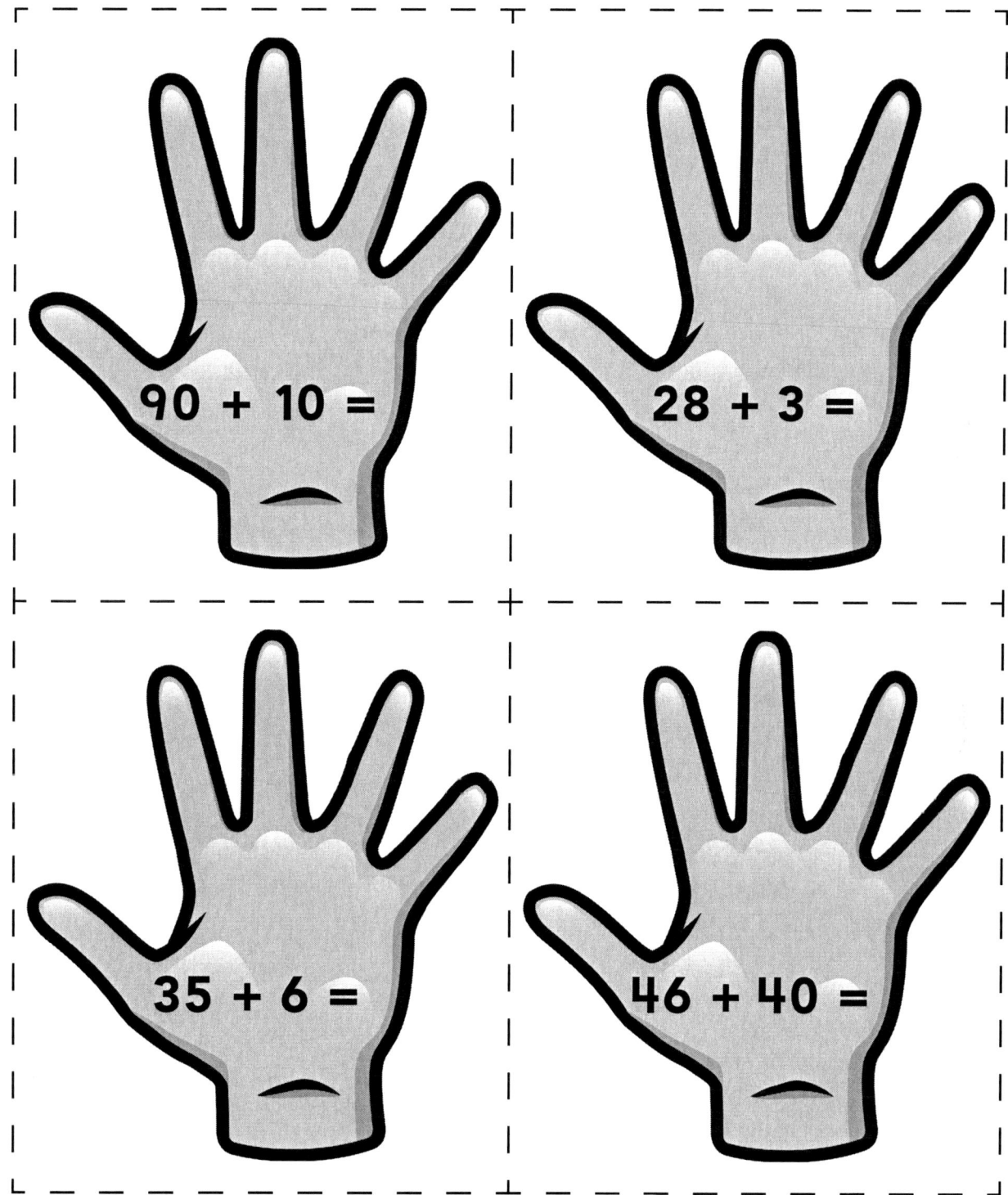

90 + 10 =

28 + 3 =

35 + 6 =

46 + 40 =

Handy Math
Equation Cards (cont.)

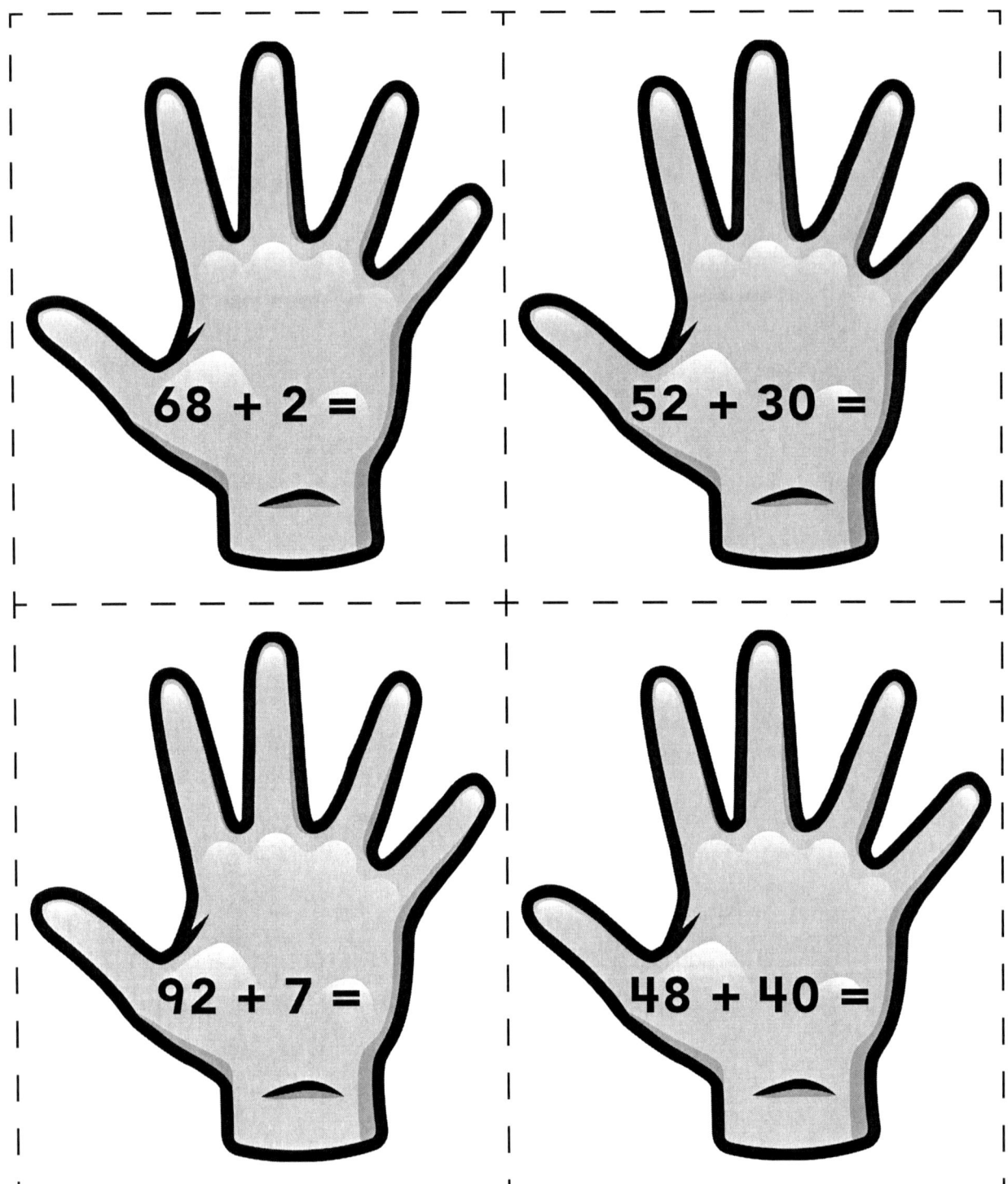

- 68 + 2 =
- 52 + 30 =
- 92 + 7 =
- 48 + 40 =

Handy Math
Equation Cards (cont.)

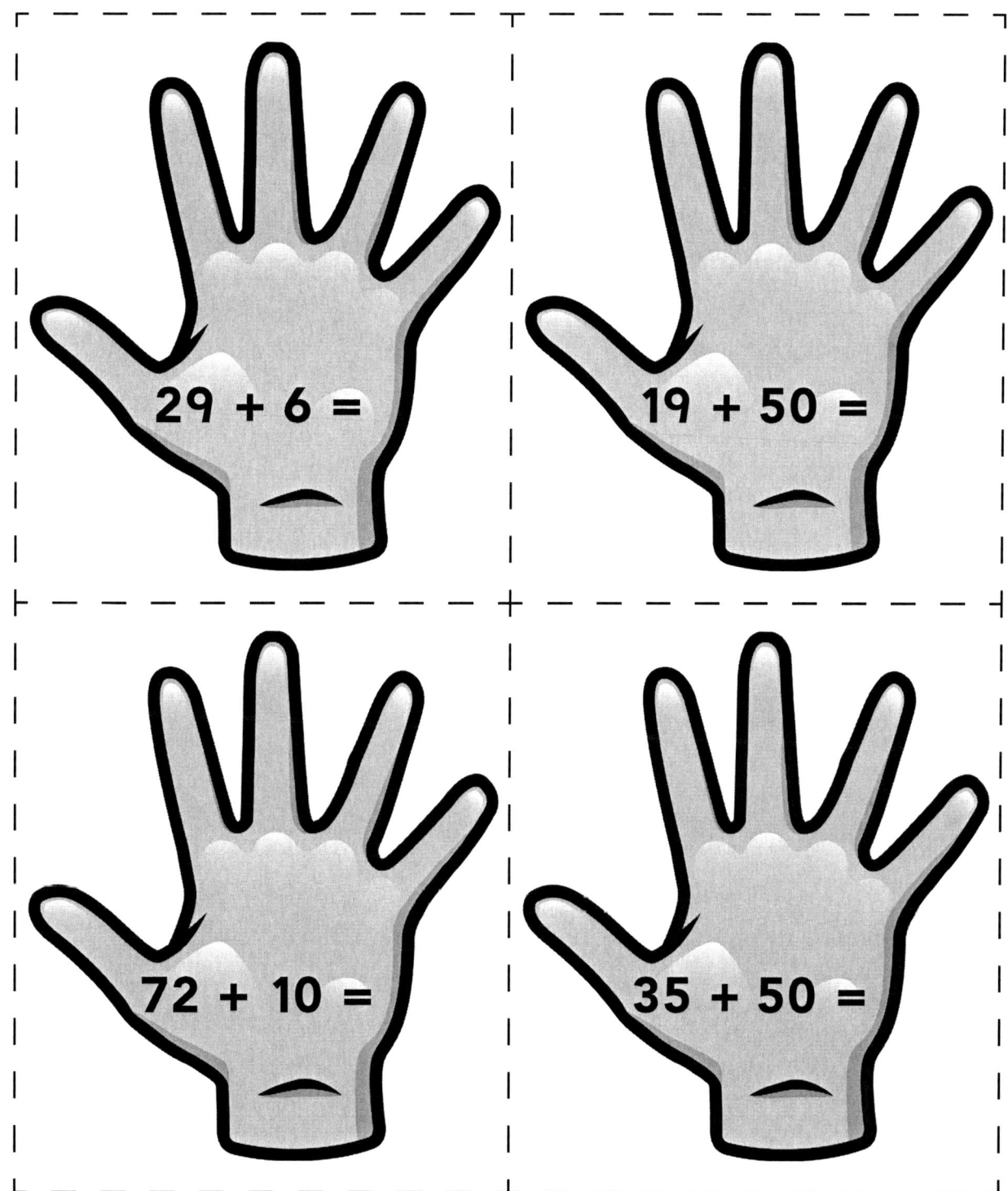

29 + 6 =

19 + 50 =

72 + 10 =

35 + 50 =

Handy Math
Equation Cards (cont.)

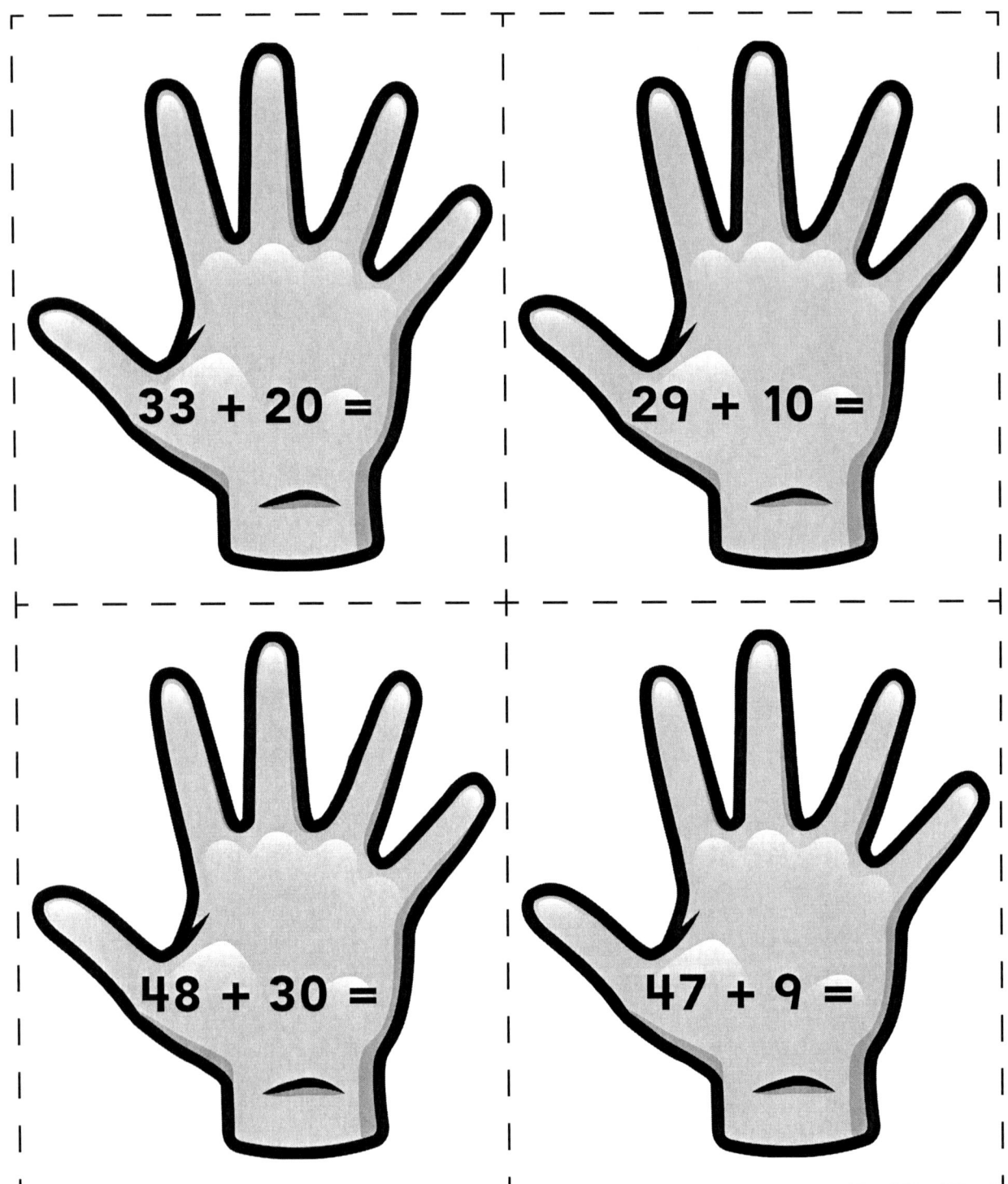

Handy Math
Equation Cards (cont.)

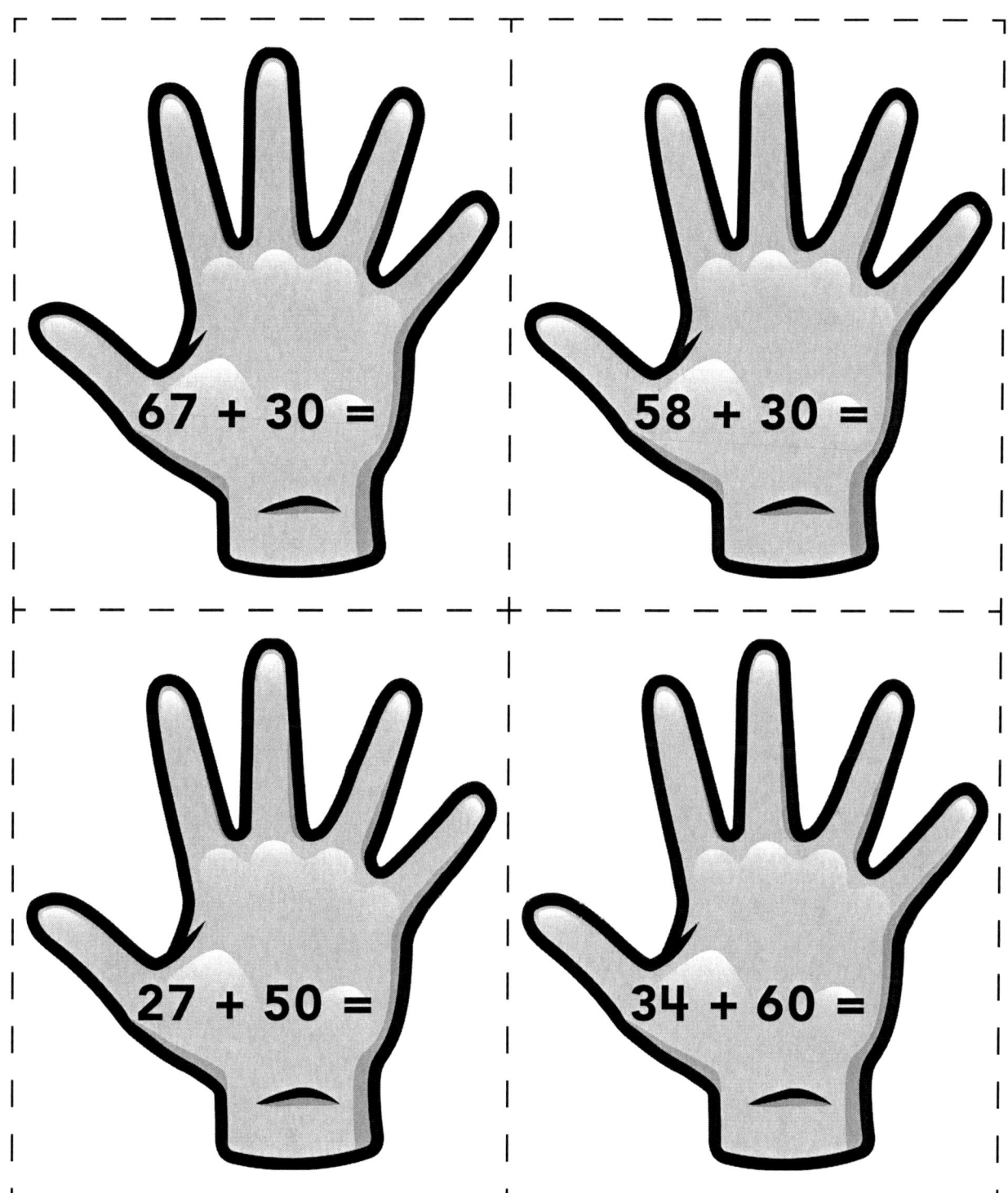

- 67 + 30 =
- 58 + 30 =
- 27 + 50 =
- 34 + 60 =

Handy Math
Equation Cards *(cont.)*

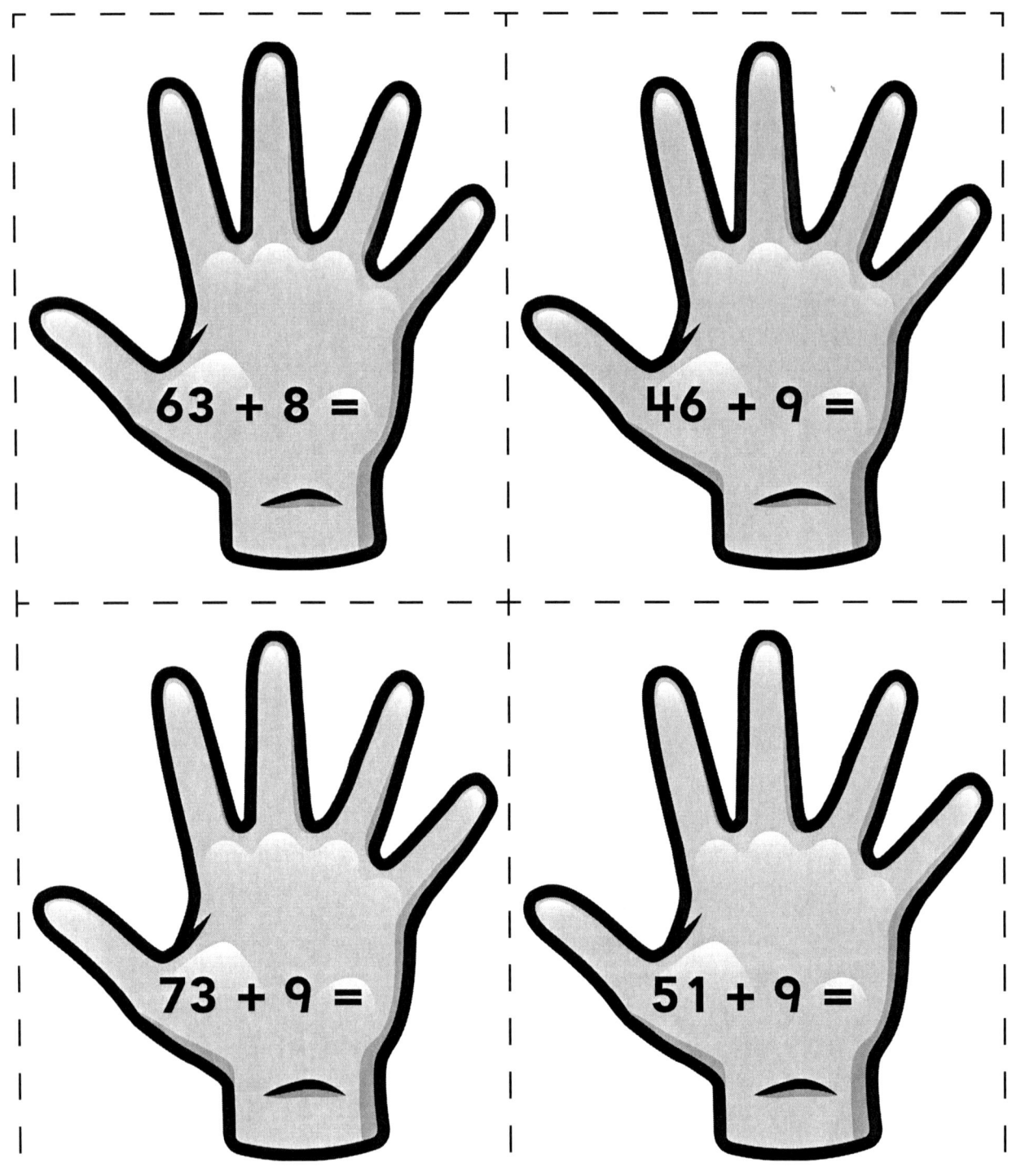

Ten: More or Less

Domain

Numbers and Operations in Base Ten

Standard

Given a 2-digit number, mentally find 10 more or 10 less than the number without having to count; explain the reasoning used.

Number of Players

2 Players

Materials

- *Ten: More or Less Spinner* (page 89)
- *Ten: More or Less Recording Page* (page 90)
- number cubes

GET PREPARED!

- Copy and cut out a *Ten: More or Less Spinner* for each pair of players.
- Make one copy of the *Ten: More or Less Recording Page* for each player.
- Collect two number cubes for each pair of players.

Game Directions

1. Distribute materials to players.

2. Players take turns rolling a number cube. The player who rolls the higher number is Player 1.

3. Player 1 rolls two number cubes, one after the other, and records the numbers on the *Ten: More or Less Recording Page*.

4. The first number cube rolled represents tens, and the second number cube represents ones. For example, if the first number cube rolled is a 6 and the second number cube is a 3, the number created is 63. Player 1 records the 2-digit number on the recording page.

Ten: More or Less (cont.)

5. Player 2 flicks the paperclip around the pencil in the center of the *Ten: More or Less Spinner* to determine if Player 1 calculates a number that is 10 more or 10 less than the number he or she rolled.

6. Player 1 calculates and records the new number that is either 10 more or 10 less than the number he or she rolled.

7. If correct, Player 2 places a check mark in the correct column on the recording page.

8. Player 2 rolls two number cubes, repeating steps 3 to 7, with Player 1 checking the work for Player 2.

9. Players continue for 5 rounds. The player with more check marks wins!

Ten: More or Less
Spinner

Directions: Copy and cut out one spinner for each pair of players. For steps on how to assemble this spinner, see page 10.

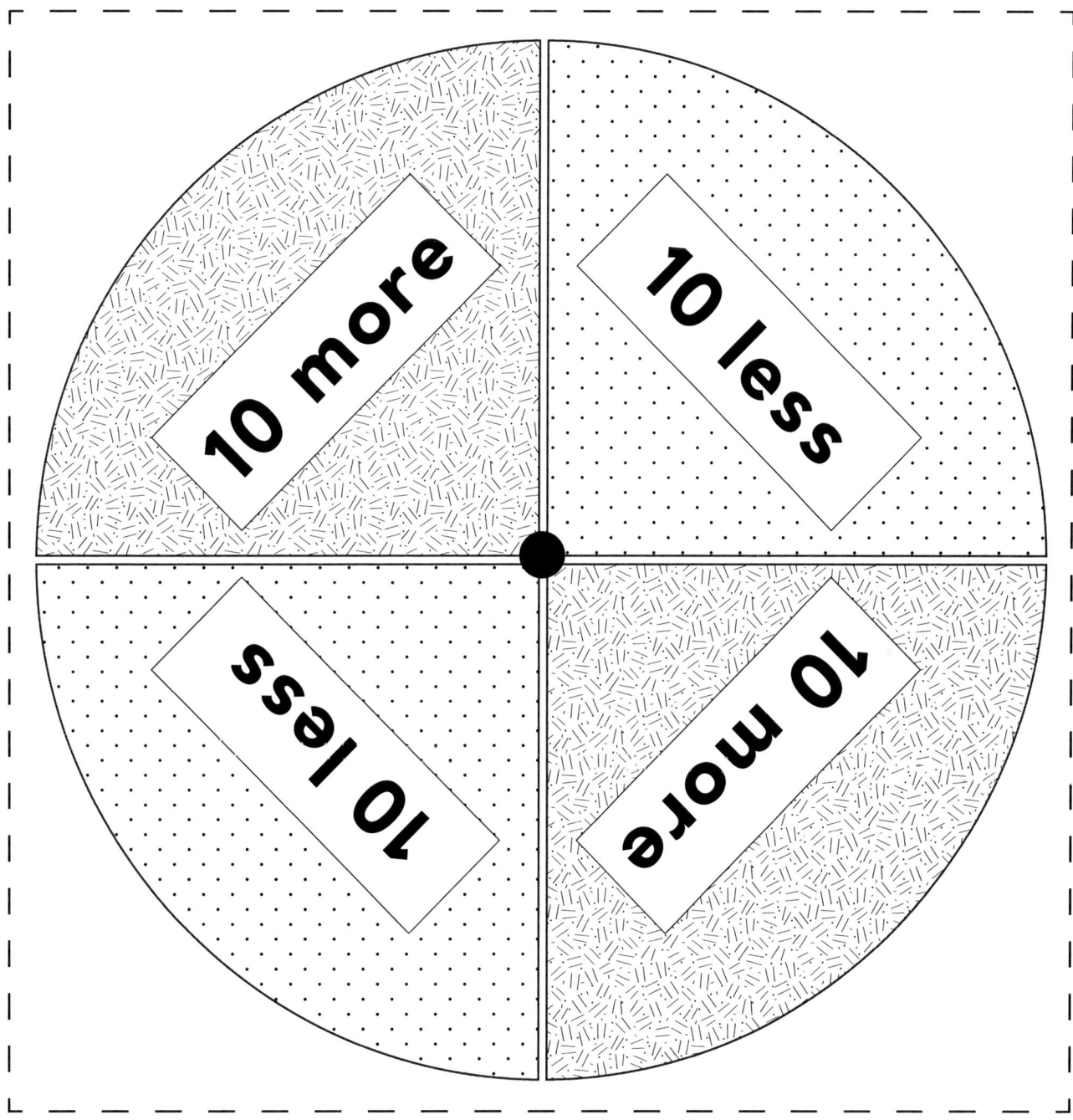

Numbers and Operations in Base Ten

Ten: More or Less
Recording Page

Directions: Use this page to record as you play the game.

Player Name: _____

	1	2	3	4	5
Cube 1					
Cube 2					
2-Digit Number					
10 More or 10 Less?					
New Number					
Check					

Numbers and Operations in Base Ten

Multiples Subtraction

Domain
Numbers and Operations in Base Ten

Standard
Subtract multiples of 10 in the range 10–90 from multiples of 10 in the range of 10–90 (positive or 0 differences), using concrete models or drawings and strategies based on place value, properties of operations, and/or the relationship between addition and subtraction; relate the strategy to a written method and explain the reasoning used.

Number of Players
2 Players

Materials
- *Multiples Subtraction Cards* (page 92)
- *Multiples Subtraction Game Board* (page 93)
- concrete models such as base ten blocks or cubes
- drawing paper

GET PREPARED
- Copy and cut out a set of *Multiples Subtraction Cards* for each pair of players.
- Copy and cut out a *Multiples Subtraction Game Board* for each player.
- Collect a set of concrete models or base ten blocks or cubes and drawing paper for each pair of players.

Game Directions

1. Players shuffle a set of *Multiples Subtraction Cards* and place the stack facedown between them.

2. Player 1 selects a card and answers the subtraction problem. If Player 1 is correct, he or she covers the correct answer on his or her *Multiples Subtraction Game Board* with the card. If a card is selected with an answer already covered or if the answer is incorrect, the card is placed at the bottom of the stack. The player loses that turn.

3. Players may use base ten blocks or other concrete models or may use drawings to explain the reasoning used for each calculation.

4. Player 2 selects a card and follows steps 2 and 3. The first player to fill his or her board wins!

Multiples Subtraction
Cards

Directions: Copy and cut out a set of cards for each pair of players.

90 − 50	70 − 40	90 − 40
70 − 50	40 − 30	60 − 20
50 − 20	70 − 20	60 − 40
30 − 10	80 − 70	90 − 80

Multiples Subtraction
Game Board

Directions: Copy and cut out a game board for each player.

Multiples Subtraction	10	20	30	40	50
Multiples Subtraction	10	20	30	40	50
Multiples Subtraction	10	20	30	40	50

Simon Says, Compare Me

Domain

Measurement and Data

Standard

Order three objects by length; compare the lengths of two objects indirectly by using a third object.

Number of Players

4 to 5 players

Materials

- a class quantity of items of different lengths (such as 25 paper clips, 25 same-size scissors, and 25 new pencils)
- *Simon Says, Compare Me Cards* (page 95)

GET PREPARED!

- Copy and cut out one set of *Simon Says, Compare Me Cards* for each group of players.
- Collect a set of the same three objects of different lengths for each player (such as one small paper clip, one pair of scissors, and one new pencil for each group of players).

Game Directions

1. Distribute materials to players and a leader is chosen for each group.

2. Players explore the objects, placing them side-by-side, and discussing the relative length of the objects (i.e., longer, shorter).

3. All players stand at their desks, facing the leader.

4. The leader begins the game of "Simon Says" by selecting and reading aloud a *Simon Says, Compare Me Card*. For example, "Simon says hold up the shortest item." All players pick up the shortest item and hold it up.

5. If a player holds up an incorrect item, the leader asks the player to sit down. If the card read by the leader does not begin with Simon says, any players who hold up an object are asked to sit down.

6. All items are set back down before another card is selected.

7. The last player standing is the winner of the game.

Measurement and Data

Simon Says,
Compare Me Cards

Directions: Copy and cut out one set of cards.

Simon says, "Hold up the longest item."	Hold up the shortest item.	Simon says, "Hold up the shortest item."	Simon says, "Hold up an item that is shorter than the _____."
Hold up an item that is shorter than the ____.	Simon says, "Hold up an item that is longer than the ____."	Simon says, "Hold up the longest item in your left hand."	Simon says, "Hold up the shortest item in your right hand."
Simon says, "Order your three objects from shortest to longest."	Order your three objects from shortest to longest.	Simon says, "Order your three objects from longest to shortest."	Order your three objects from longest to shortest.

Measurement and Data

Measurement, Measurement, Measurement

Domain

Measurement and Data

Standard

Express the length of an object as a whole number of length units by laying multiple copies of a shorter object (the length unit) end to end; understand that the length measurement of an object is the number of same-size length units that span it with no gaps or overlaps. Limit to contexts where the object being measured is spanned by a whole number of length units with no gap or overlaps.

Number of Players

3 Players

Materials

- *Measurement, Measurement, Measurement Cards* (pages 98–99)
- objects for players to measure (e.g., pencils, books, shoes, index cards)
- objects to use as units of measure (e.g., paper clips, cubes, erasers, bear counters)
- number cubes
- paper bag

GET PREPARED!

- Copy and cut out one set of *Measurement, Measurement, Measurement Cards* for each group of players.
- Place three items to be measured in a paper bag for each group of players.
- Collect a handful of each of three different types of items to use as units of measure and a number cube for each group of players.
- Prepare one example with a unit of measure and a corresponding question for players.
- Fill in the blanks on the *Measurement, Measurement, Measurement Cards* with the names of the items you choose to use.

Measurement, Measurement, Measurement (cont.)

Game Directions

1. Distribute material to players.

2. Each player rolls the number cube to determine who will be the leader. The leader changes each round and is determined by the player who rolls the highest number at the start of every round.

3. Players place the bag of objects to be measured and the objects to be used as measurement units between them.

4. The leader shuffles the cards and places them facedown. The top card is drawn and read aloud by the leader.

5. Player 1 follows the directions to measure an object with the indicated units. For example, "Use paperclips to measure the length of the pencil. How many paperclips long is the pencil?" Player 1 then answers the question on the card using the correct units of measure.

6. Player 2 checks the work of Player 1. If Player 1 is correct, he or she places the card in his or her "winning pile." If Player 1 answers incorrectly, the card is placed in the "discard pile."

7. The leader repeats the procedure, selecting another card and reading it out loud for the next player.

8. After 6 rounds, the player with the most cards in his or her pile wins!

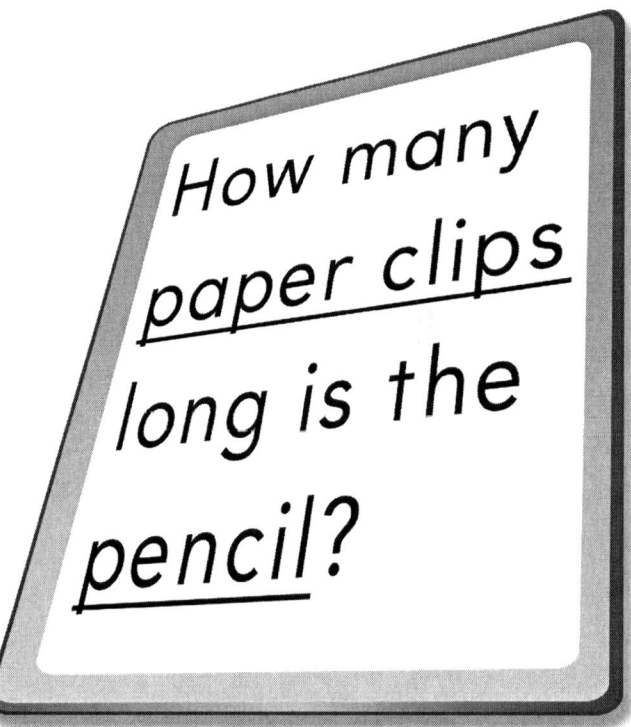

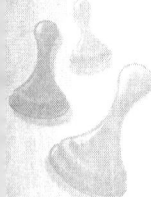

Measurement and Data

Measurement, Measurement, Measurement
Cards

Directions: Copy and cut out two sets of cards for each group of players. Fill in the blanks on each card with one of the objects you have provided for players to measure and a unit of measurement.

Use _____ to measure the length of the _____.
How many _____ long is the _____?

Use _____ to measure the length of the _____.
How many _____ long is the _____?

Use _____ to measure the length of the _____.
How many _____ long is the _____?

Use _____ to measure the length of the _____.
How many _____ long is the _____?

Measurement and Data

Measurement, Measurement, Measurement
Cards (cont.)

Use _____ to measure the length of the _____.
How many _____ long is the _____?

Use _____ to measure the length of the _____.
How many _____ long is the _____?

Use _____ to measure the length of the _____.
How many _____ long is the _____?

Use _____ to measure the length of the _____.
How many _____ long is the _____?

Use _____ to measure the length of the _____.
How many _____ long is the _____?

It's the Right Time

Domain
Measurement and Data

Standard
Tell and write time in hours and half-hours, using analog and digital clocks.

Number of Players
2 Players

Materials
- *It's the Right Time Spinner* (page 101)
- *It's the Right Time Game Board* (pages 102–103)
- *It's the Right Time Recording Sheet* (page 104)
- game pieces for each player (e.g., two-color counters, small colored cubes)
- paper clip and pencil

GET PREPARED!
- Copy one *It's the Right Time Recording Sheet* for each pair of players.
- Copy and cut out *It's the Right Time Spinner* and one *It's the Right Time Game Board* for each pair of players.
- Collect two game pieces, a paper clip, and a pencil for each pair of players.

Game Directions

1. Players take turns flicking the paper clip on the spinner. The player who spins the higher number is Player 1.

2. Players place their game pieces at the "Start" of the game board.

3. Player 1 spins and moves his or her game board piece the number of spaces shown.

4. To stay on the space, Player 1 must correctly tell the time shown on the clock and record the digital time on his or her *It's the Right Time Recording Sheet*. If Player 1 is incorrect, he or she must move back one space.

5. Player 2 spins and repeats steps 3 and 4.

6. The first player to reach the "Finish" space or beyond wins.

Measurement and Data

It's the Right Time
Spinner

Directions: Copy and cut out a spinner for each pair of players. For steps on how to assemble this spinner, see page 10.

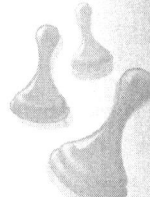

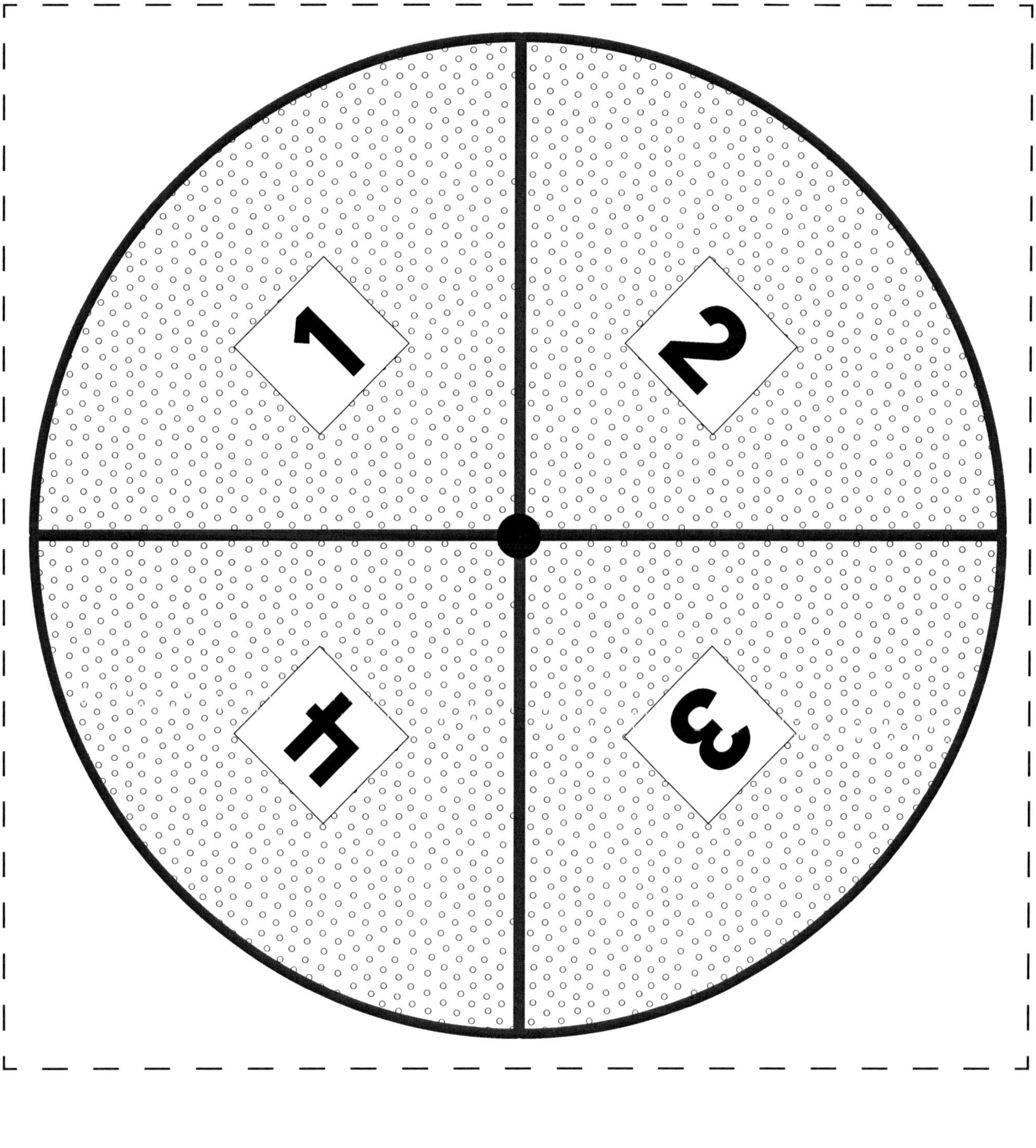

Measurement and Data

It's the Right Time
Game Board

Directions: Copy and cut out the game board. Tape it to the game board on page 103.

Measurement and Data

It's the Right Time
Game Board (cont.)

Start

Five-thirty

Two-thirty

Right time

4:30

Ten O'Clock

tape here

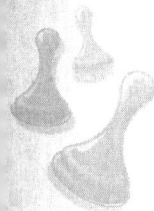

Measurement and Data

It's the Right Time
Recording Sheet

Directions: Record the digital times shown on the clock.

Player 1	Player 2

Measurement and Data

Graph It!

Domain
Measurement and Data

Standard
Organize, represent, and interpret data with up to three categories; ask and answer questions about the total number of data points, how many in each category, and how many more or less are in one category than in another.

Number of Players
2 Players

Materials
- *Graph It! Graphs* (pages 106–110)
- *Graph It! Graph Cards* (pages 111–120)
- number cube

GET PREPARED!
- Copy and cut out one set of *Graph It! Graph Cards* and one set of *Graph It! Graphs* for each pair of players.
- Collect one number cube for each pair of players.

Game Directions

1. Distribute materials to players.

2. Players take turns rolling the number cube. The player who rolls the lower number is Player 1. Players shuffle the game cards and place them facedown in a stack between them.

3. Player 1 draws a card, reads the question, and refers to the graph to provide an answer.

4. If Player 2 determines that Player 1 has answered the question correctly, then Player 1 keeps the card in his or her winning pile. If incorrect, Player 2 has the opportunity to answer the question. If correct, he or she may keep the card. If incorrect, the card is placed in the discard pile.

5. Player 2 draws and repeats steps 3 and 4.

6. When all the cards have been used, the player with more cards in his or her winning pile wins.

7. Follow the same steps using the other graphs and corresponding cards. Determine how many different games players should play.

© Shell Education #51288—Math Games: Skill-Based Practice **105**

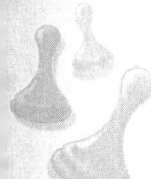

Measurement and Data

Graph It!
Graph 1

Directions: Copy and cut out the graph for each pair of players.

Eye Colors in Our Class

Number of Students	Brown	Blue	Green
10	■		
9	■		
8	■		
7	■	■	
6	■	■	
5	■	■	
4	■	■	■
3	■	■	■
2	■	■	■
1	■	■	■

Eye Color

106   #51288—Math Games: Skill-Based Practice   © Shell Education

Measurement and Data

Graph It!
Graph 2

Directions: Copy and cut out the graph for each pair of students.

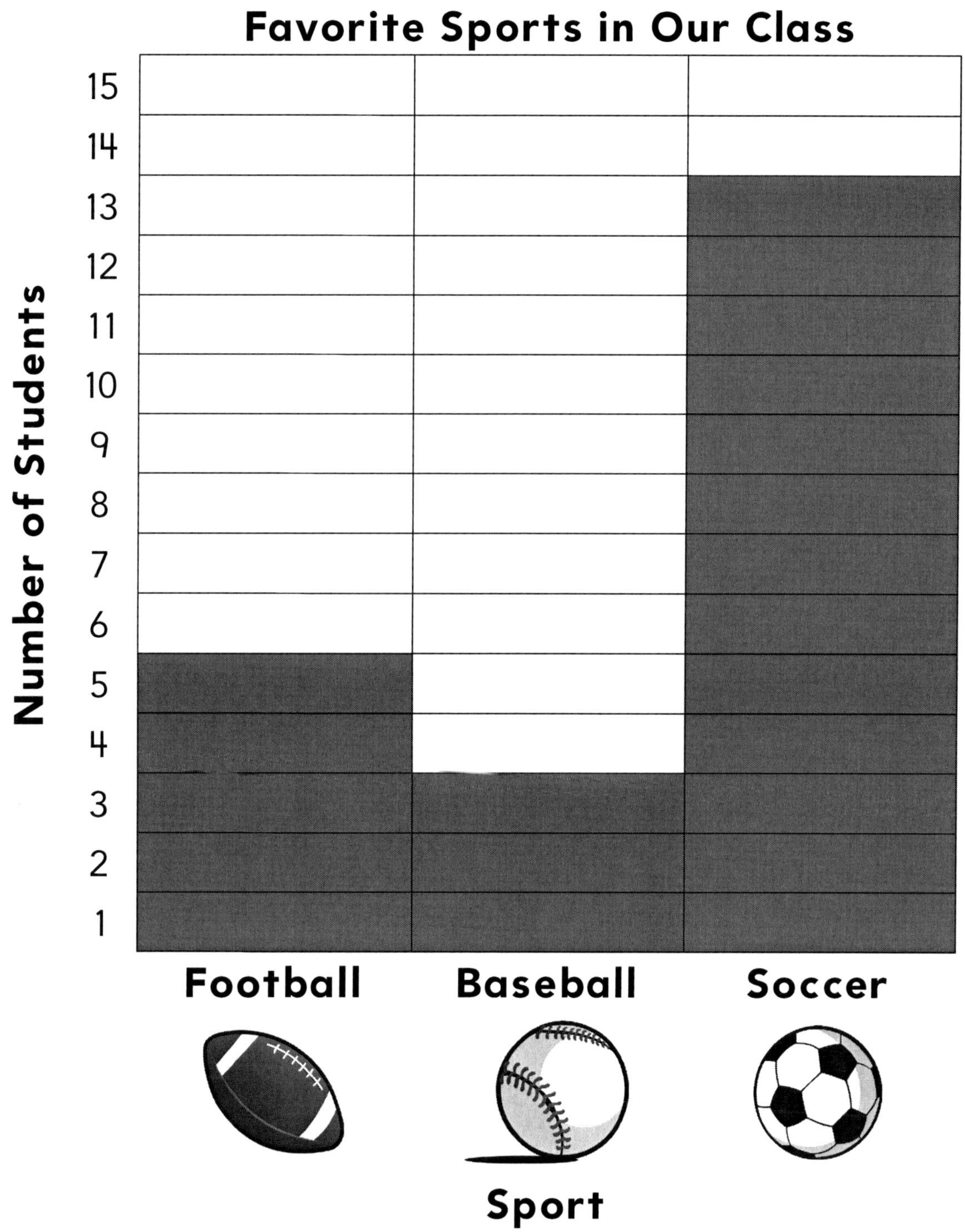

Favorite Sports in Our Class

(Bar graph — Number of Students vs. Sport)
- Football: 5
- Baseball: 3
- Soccer: 13

Measurement and Data

Graph It!
Graph 3

Directions: Copy and cut out the graph for each pair of players.

Favorite Ice Cream Flavors in Our Class	
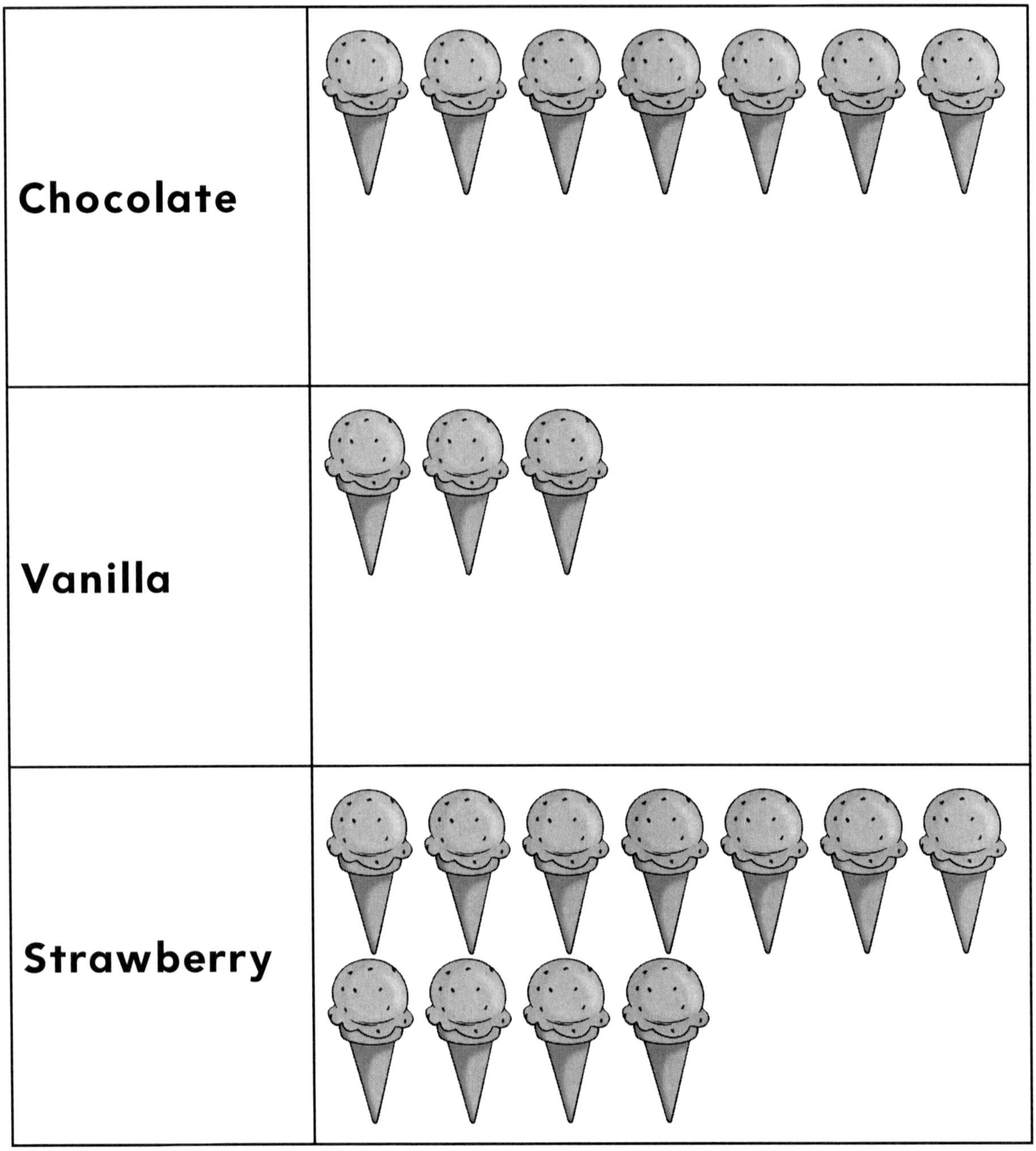 Chocolate	🍦🍦🍦🍦🍦🍦🍦
Vanilla	🍦🍦🍦
Strawberry	🍦🍦🍦🍦🍦🍦🍦 🍦🍦🍦🍦

Measurement and Data

Graph It!
Graph 4

Directions: Copy and cut out the graph for each pair of players.

Favorite Foods in Our Class

Pizza	8 pizzas
Spaghetti	5 plates
Hamburger	9 hamburgers

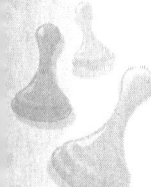

Measurement and Data

Graph It!
Graph 5

Directions: Copy and cut out the graph for each pair of players.

Ways We Get to School

Walk	(10 walkers)
Bus	(5 buses)
Car	(9 cars)

110 #51288—Math Games: Skill-Based Practice © Shell Education

Measurement and Data

Graph It!
Graph 1 Cards

Directions: Copy and cut apart one set of cards for each pair of players.

How many students have brown eyes? 	How many students have blue eyes? 
How many students have green eyes? 	How many more students have brown eyes than blue eyes? 

Measurement and Data

Graph It!
Graph 1 Cards (cont.)

How many more students have brown eyes than green eyes? 	Most students have which eye color? 
The least number of students have which eye color? 	How many fewer students have green eyes than blue eyes? 

Measurement and Data

Graph It!
Graph 2 Cards

Directions: Copy and cut apart one set of cards for each pair of players.

How many students say that football is their favorite sport? 	How many students say that baseball is their favorite sport? 
How many students say that soccer is their favorite sport? 	How many more students prefer soccer over baseball?  

Measurement and Data

Graph It!
Graph 2 Cards (cont.)

How many more students prefer football over baseball?  	Which sport is the least favorite of students?   
How many fewer students prefer baseball over soccer?  	How many fewer students prefer football over soccer?  

Measurement and Data

Graph It!
Graph 3 Cards

Directions: Copy and cut apart one set of cards for each pair of players.

How many students say their favorite ice cream flavor is strawberry? 	How many students say their favorite ice cream flavor is chocolate? 
How many students say their favorite ice cream flavor is vanilla? 	How many more students prefer strawberry ice cream over vanilla ice cream? 

© Shell Education #51288—Math Games: Skill-Based Practice 115

Measurement and Data

Graph It!
Graph 3 Cards (cont.)

How many more students prefer strawberry ice cream over chocolate ice cream? 	Which ice cream flavor is the least favorite of students? 
How many fewer students prefer vanilla ice cream than chocolate ice cream? 	How many fewer students prefer vanilla ice cream than strawberry ice cream? 

Measurement and Data

Graph It!
Graph 4 Cards

Directions: Copy and cut apart one set of cards for each pair of players.

How many students say their favorite food is a hamburger? 	How many students say their favorite food is pizza? 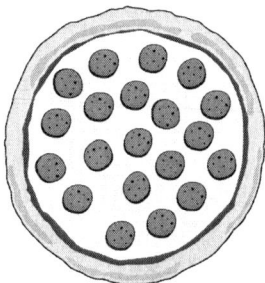
How many students say their favorite food is spaghetti? 	How many more students prefer a hamburger over pizza? 

Measurement and Data

Graph It!
Graph 4 Cards (cont.)

How many more students prefer hamburgers over spaghetti?

Which food is the least favorite of students?

How many fewer students prefer spaghetti over hamburgers?

How many fewer students prefer spaghetti over pizza?

Measurement and Data

Graph It!
Graph 5 Cards

Directions: Copy and cut apart one set of cards for each pair of players.

How many students are driven in a car to school? 	How many students take the bus to school? 
How many students walk to school? 	How many more students are driven to school in a car than take the bus to school? 

Measurement and Data

Graph It
Game 5 Cards (cont.)

How many more students are driven in a car to school than walk to school? 	Which type of transportation is the least used by students? 
How many fewer students take the bus than walk to school? 	How many fewer students take the bus than are driven in a car to school? 

Tell Me About It Geometry

Domain
Geometry

Standard
Distinguish between defining attributes (e.g., triangles are closed and three-sided vs. non-defining attributes (e.g., color, orientation, overall size); build and draw shapes to possess defining attributes.

Number of Players
3 Players

Materials
- Tell Me About It Geometry Cards (pages 122–123)
- Tell Me About It Tally Sheet (page 124)
- whiteboards
- dry-erase markers
- number cubes

GET PREPARED!
- Copy and cut out one set of the *Tell Me About It Geometry Cards* for each group of players.
- Copy a *Tell Me About It Tally Sheet* for each group of players.

Game Directions

1. Distribute materials to players and designate a leader.

2. Players roll the number cube to determine who goes first. The player with the highest number is Player 1. The player with the second highest number is Player 2. If there are three players, then the player with the lowest number is Player 3.

3. The leader draws a *Tell Me About It Geometry Card* and reads it to Player 1.

4. Player 1 draws on an individual whiteboard or points to objects in the room to describe the answer.

5. If correct, Player 1 is awarded one point. If incorrect, Player 2 has the opportunity to answer the question and receive a point for a correct answer. Players record points on their tally sheet.

6. Players rotate roles. The leader becomes Player 1, Player 1 becomes Player 2, and Player 2 becomes the leader. Players 1 and 2 repeat steps 4 and 5.

7. Once all the questions have been answered, the player with the most points wins!

Geometry

Tell Me About It
Geometry Cards

Directions: Copy and cut out one set of cards for each group of players.

My shape is a solid with 2 circular faces connected by a curved surface. It looks like a can. Draw me or point to me in our classroom.	My shape is round. I have one side. Draw me or point to me in our classroom.
My shape is a closed figure with 3 sides and 3 angles. Draw me or point to me in our classroom.	My shape is a 3D figure with 6 square faces. Draw me or point to me in our classroom.
My shape is a round object with a curved surface. Draw me or point to me in our classroom.	My shape has only 2 lines that are parallel.

Tell Me About It
Geometry Cards (cont.)

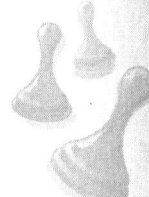

My shape is a solid pointed figure that has a flat round base.

My shape is a 6-sided figure with 6 angles.

Tell me one thing that is different between a circle and a sphere.

Tell me one thing that is the same and one thing that is different between a cube and a triangle.

Tell me one thing that is the same about a rectangle and a square.

Tell me one thing that is different about a rectangle and a square.

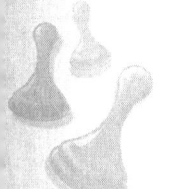

Geometry

Tell Me About It
Tally Sheet

Directions: Copy and cut out one tally sheet for each group of three players.

Player 1	Player 2	Player 3

Player 1	Player 2	Player 3

Player 1	Player 2	Player 3

Shape-Maker

Domain
Geometry

Standard
Compose 2-dimensional shapes (rectangles, squares, trapezoids, triangles, half circles, and quarter-circles) or 3-dimensional shapes (cubes, right rectangular prisms, right circular cones, and right circular cylinders) to create a composite shape, and compose new shapes from the composite shape.

Number of Players
2 Players

Materials
- *Pattern Block Set* (page 127)
- *Shape-Maker Game Board* (pages 128–129)
- *Shape-Maker Spinner* (page 130)
- game pieces for each player (e.g., two-color counters, small colored cubes)
- paper clips and pencils
- number cubes

GET PREPARED!

- Copy and cut out one *Shape-Maker Game Board* and one *Shape-Maker Spinner* for each pair of players.
- Copy and cut out one set of the *Pattern Block Set* sheets for each pair of players.
- Collect two game pieces, a pencil, and a paper clip for each pair of players.

Game Directions

1. Distribute materials to players.

2. Players roll the number cube to determine who goes first. The player with the higher roll is Player 1.

3. Each player places his or her game piece at the "Start" of the game board.

Geometry

Shape-Maker (cont.)

4. Player 1 flicks the paperclip around the pencil in the center of the spinner and moves forward on the game board to the next space of the shape spun.

5. Player 1 creates the shape that was spun using more than one pattern block without overlapping pieces or gaps. For example, if Player 1 lands on a square, he or she picks up four squares from the pattern blocks set to create a large square, as shown below. If player 2 lands on a trapezoid, he or she can use three triangles to create that shape, as shown below.

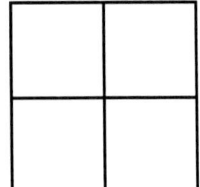

6. Player 2 repeats steps 4 and 5.

7. The first player to reach the end of the game board wins!

Geometry

Pattern Block Set

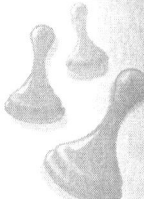

Directions: Copy and cut out one set of the pattern blocks for each pair of players.

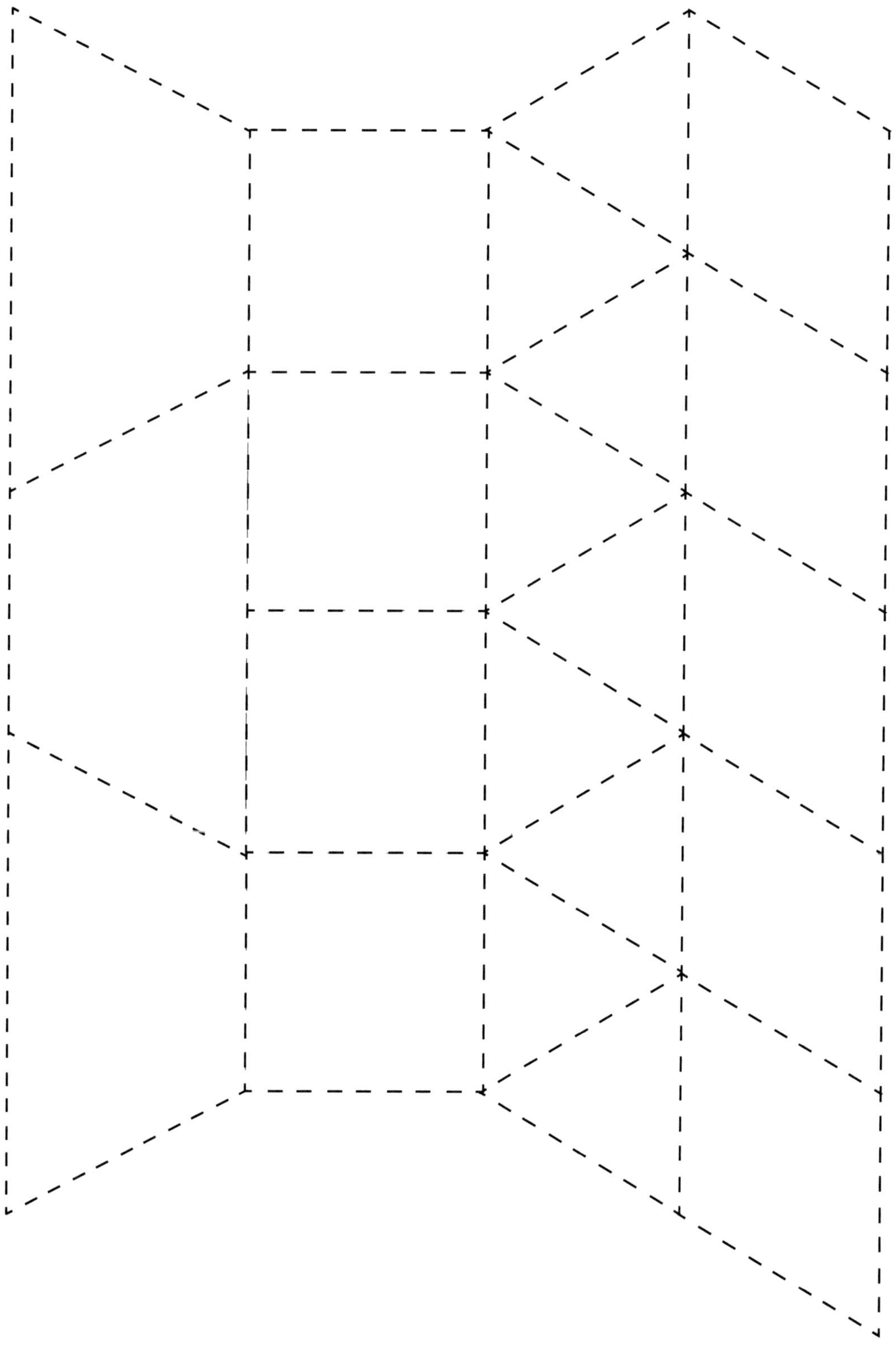

© Shell Education #51288—Math Games: Skill-Based Practice

Geometry

Shape-Maker
Game Board

Directions: Copy and cut out the game board. Tape it to the game board on page 129.

128 #51288—Math Games: Skill-Based Practice © Shell Education

Geometry

Shape-Maker
Game Board (cont.)

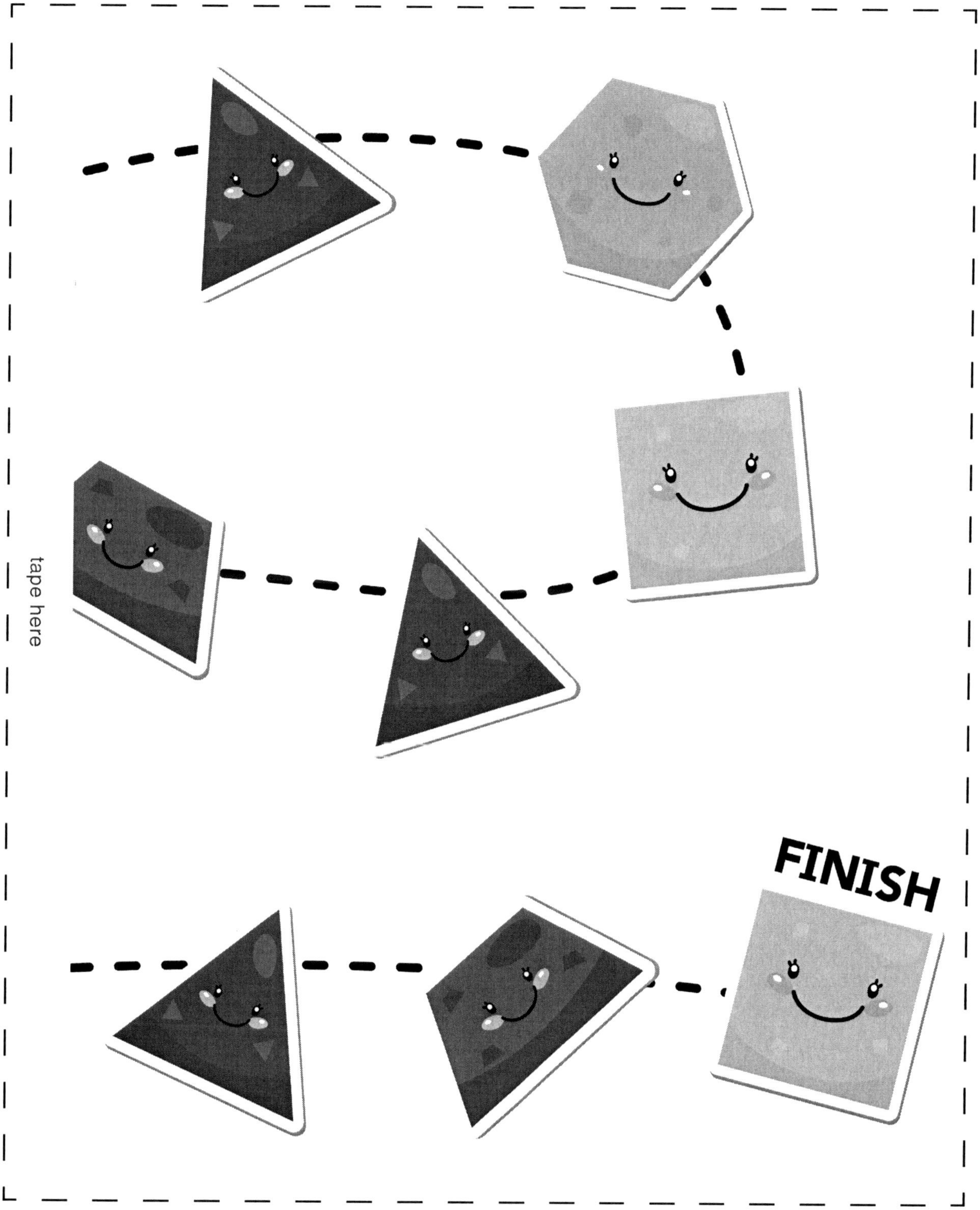

tape here

FINISH

© Shell Education

Geometry

Shape-Maker
Spinner

Directions: Copy and cut out one spinner per pair of players. For steps on how to assemble this spinner, see page 10.

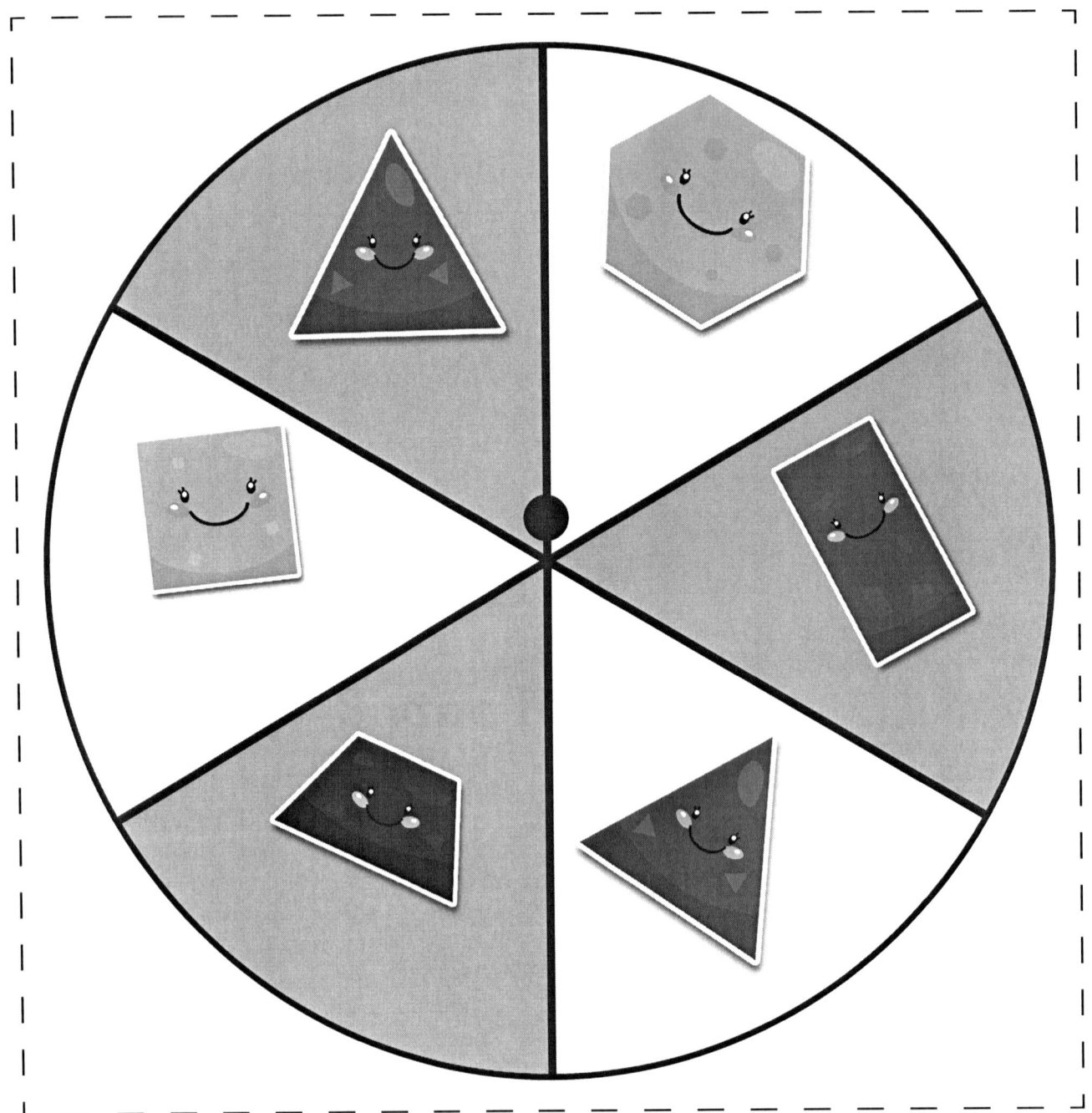

Geometry

Sharing Shapes

Domain
Geometry

Standard
Partition circles and rectangles into two and four equal shares; describe the shapes using the words *halves*, *fourths*, and *quarters*; and use the phrases *half of*, *fourth of*, and *quarter of*. Describe the whole as two of or four of the shares. Understand for these examples that decomposing into more equal shares creates smaller shares.

Number of Players
2 Players

Materials
- *Sharing Shapes Spinner* (page 133)
- *Sharing Shapes Game Board* (pages 134–136)
- *Sharing Shapes Sheet* (page 137)
- paper clips and pencils
- number cubes
- game piece for each player (e.g., two-color counters, small colored cubes, mini erasers)
- pencils

GET PREPARED!

- Copy and cut out one *Sharing Shapes Game Board* for each pair of players.
- Copy and cut out one *Sharing Shapes Spinner* and the *Sharing Shapes Sheet* for each pair of players.
- Collect a paper clip, a pencil, and a number cube for each pair of players.

Game Directions

1. Distibute materials to players.
2. Players place their game pieces at the "Start" of the game board.
3. Players roll the number cube to determine who goes first. The player with the lower roll is Player 1.

© Shell Education   #51288—Math Games: Skill-Based Practice   **131**

Geometry

Sharing Shapes (cont.)

4. Player 1 flicks the paperclip around the pencil in the center of the spinner and reads the words.

5. Player 1 uses the pencil to draw a line to partition any shape from their *Sharing Shapes Cards* according to the fraction displayed on the spinner. If correct, Player 1 rolls the number cube to see how many spaces he or she can move forward.

6. If Player 1 cannot divide the shape correctly, Player 2 corrects him or her and no move is made.

7. Player 2 takes a turn and repeats Steps 4 to 6.

8. The first player to reach "Finish" wins!

Sharing Shapes
Spinner

Directions: Copy and cut out one spinner for each pair of players. For steps on how to assemble this spinner, see page 10.

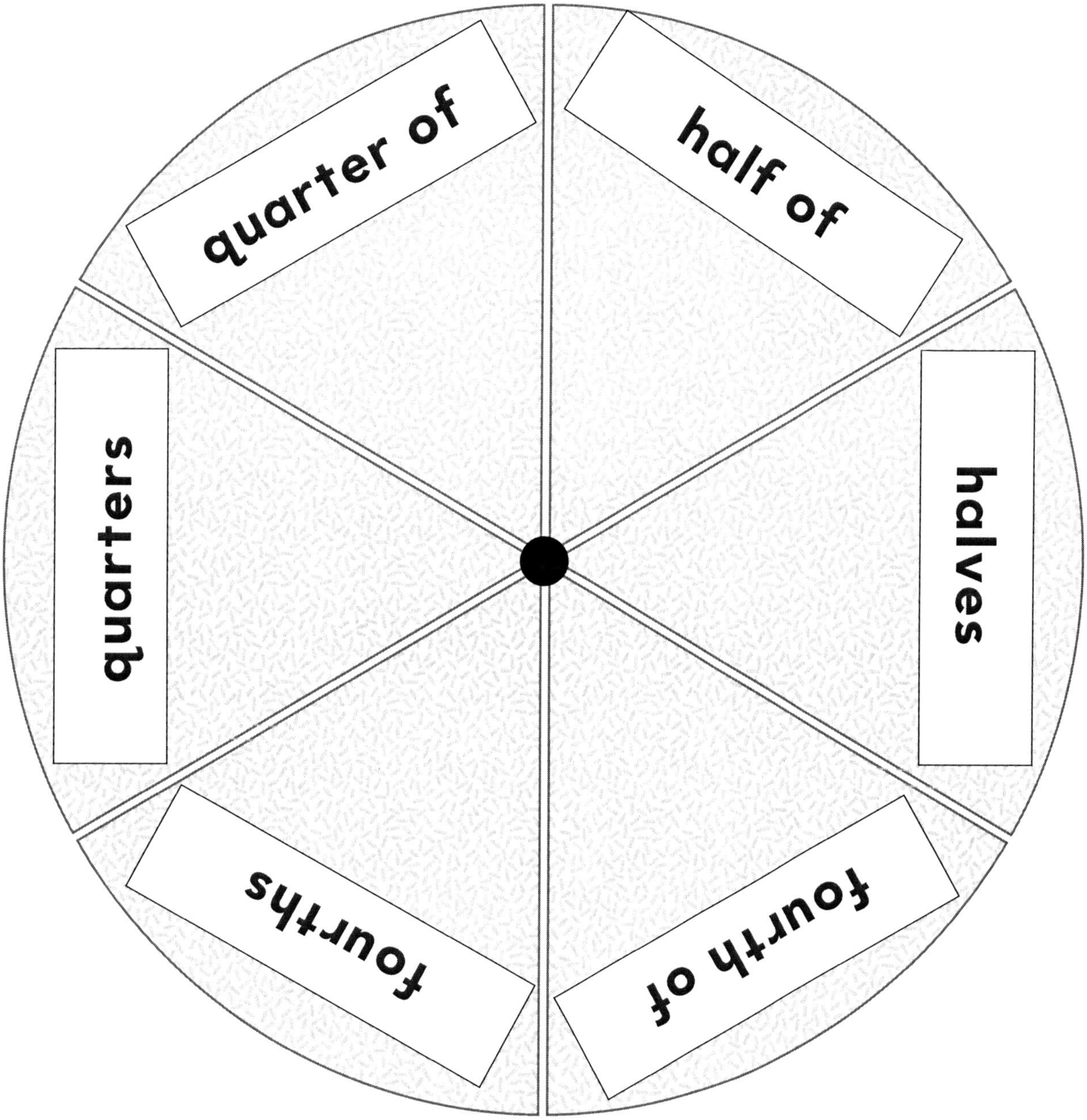

- quarter of
- half of
- halves
- fourth of
- fourths
- quarters

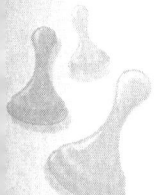

Geometry

Sharing Shapes
Game Board

Directions: Copy and cut out the game board. Tape it to the game boards on pages 135–136.

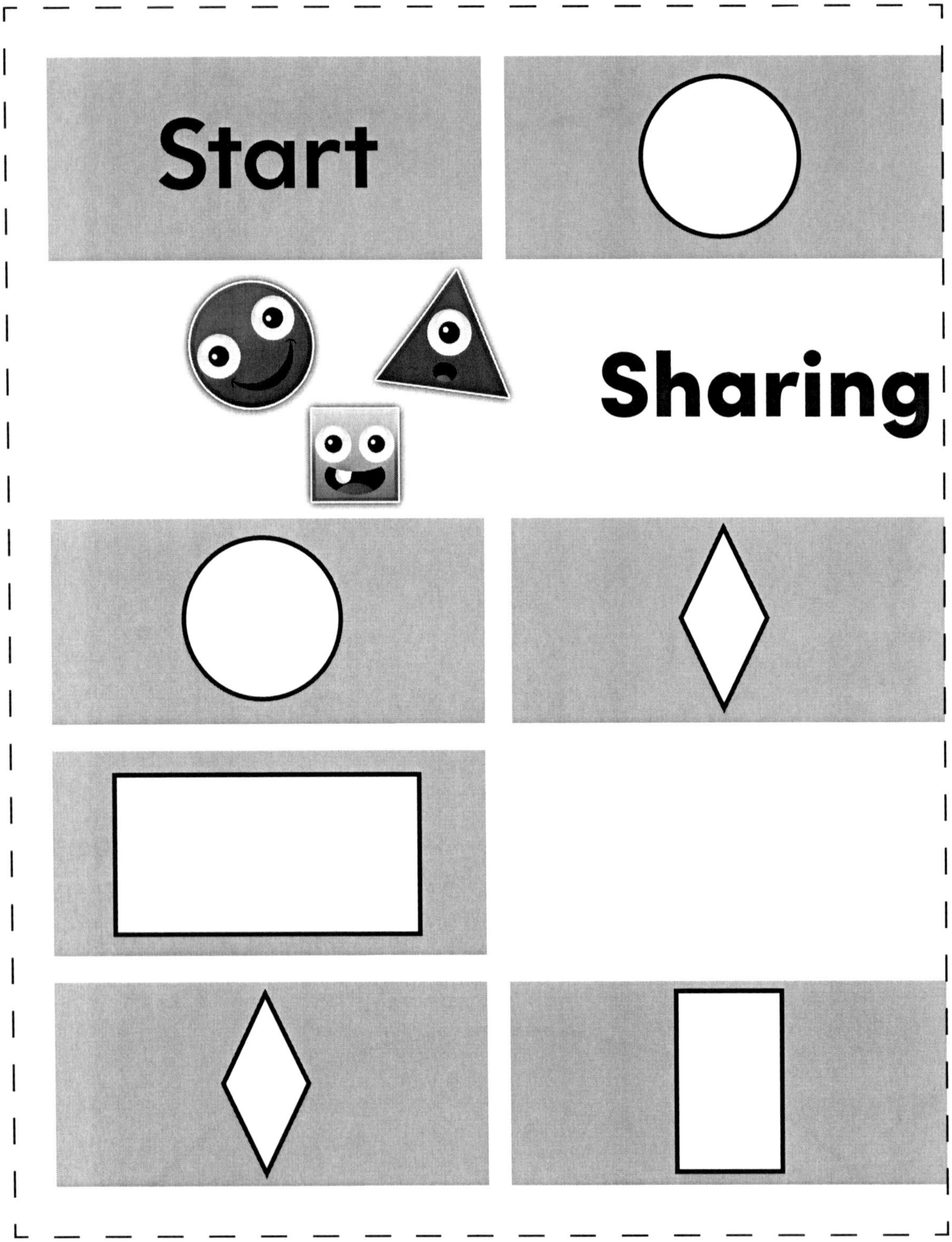

Geometry

Sharing Shapes
Game Board (cont.)

Shapes

tape here

tape here

© Shell Education   #51288—Math Games: Skill-Based Practice   135

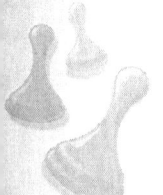

Geometry

Sharing Shapes
Game Board (cont.)

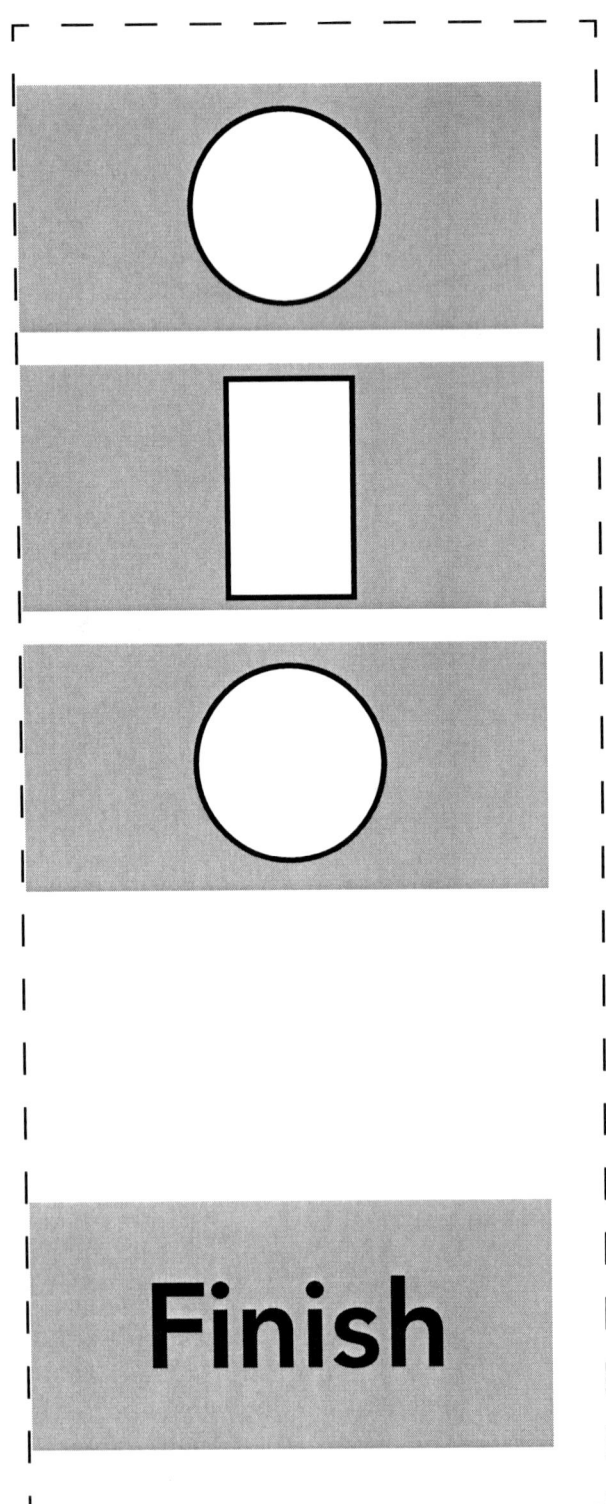

Geometry

Sharing Shapes
Sheet

Directions: Copy and cut out one sheet for each pair of players.

© Shell Education　　　#51288—Math Games: Skill-Based Practice　　137

References Cited

Burns, Marilyn. 2009. "Win-Win Math Games." *Instructor*. Reprinted March/April, http://www.mathsolutions.com/documents/winwin_mathgames.pdf.

Hull, Ted H., Ruth Harbin Miles, and Don S. Balka. 2013. *Math Games: Getting to the Core of Conceptual Understanding*. Huntington Beach, CA: Shell Education.

National Council of Teachers of Mathematics. 2000. *Principles and Standards for School Mathematics*. Reston, VA: NCTM.

National Governors Association Center for Best Practices, and Council of Chief State School Officers. 2010. "Common Core State Standards." Washington, DC: National Governors Association Center for Best Practices, Council of Chief State School Officers. Accessed September 23, 2013, http://corestandards.org/math.

National Research Council. 2001. "Adding It Up: Helping Children Learn Mathematics." Washington, DC: National Academy Press.

National Research Council. 2004. "Engaging Schools: Fostering High School Students' Motivation to Learn." Washington, DC: National Academy Press.

Contents of the Digital Resource CD

Student Resources		
Page(s)	Title	Filename
18–27	Mailbox Templates	mailbox.pdf
28–37	Envelope Game Cards	envelp.pdf
40–41	I Can Solve It! Word Problem Cards	solveitcards.pdf
44–45	Mix-Up, Match-Up Game Board	mixupboard.pdf
46	Mix-Up, Match-Up Game Cards	mixupcards1.pdf mixupcards2.pdf
49	It's a Fact! Game Board	itsafactboard.pdf
50	It's a Fact! Game Cards	itsafactcards.pdf
52–53	Speedy Counters Race Track	counterstrack.pdf
54–59	Speedy Counters Race Cards	counterscards.pdf
61	Climb to the Top Spinner	climbspinner.pdf
62–63	Climb to the Top Game Board	climbboard.pdf
65–66	True or False Cards	trueorfalse.pdf
68	Missing Number Cards	missingcards.pdf
69–71	Unknown Number Equation Cards	unknowncards.pdf
74	Place Value Spinner	valuespinner.pdf
75–76	Place Value Game Board	valueboard.pdf
78	Roll It! Spinner	rollitspinner.pdf
79	Roll It! Recording Page	rollitpage.pdf
81–86	Handy Math Equation Cards	handycards.pdf
89	Ten: More or Less Spinner	morelessspinner.pdf
90	Ten: More or Less Recording Page	morelesspage.pdf
92	Multiples Subtraction Cards	multiplescards.pdf
93	Multiples Subtraction Game Board	multiplesboard.pdf
95	Simon Says, Compare Me Cards	simonsayscards.pdf
98–99	Measurement, Measurement, Measurement Cards	measurementcard.pdf
101	It's the Right Time Spinner	timespinner.pdf
102–103	It's the Right Time Game Board	timeboard.pdf

Contents of the Digital Resource CD (cont.)

Student Resources		
Page(s)	Title	Filename
104	It's the Right Time Recording Sheet	righttime.pdf
106	Graph It! Graph 1	graphitgraph1.pdf
107	Graph It! Graph 2	graphitgraph2.pdf
108	Graph It! Graph 3	graphitgraph3.pdf
109	Graph It! Graph 4	graphitgraph4.pdf
110	Graph It! Graph 5	graphitgraph5.pdf
111–112	Graph It! Graph 1 Cards	graphitcards1.pdf
113–114	Graph It! Graph 2 Cards	graphitcards2.pdf
115–116	Graph It! Graph 3 Cards	graphitcards3.pdf
117–118	Graph It! Graph 4 Cards	graphitcards4.pdf
119–120	Graph It! Graph 5 Cards	graphitcards5.pdf
122–123	Tell Me About It Geometry Cards	tellmegeocards.pdf
124	Tell Me About It Tally Sheet	tellmetally.pdf
127	Pattern Block Set	patternblock.pdf
128–129	Shape-Maker Game Board	shapemakerboard.pdf
130	Shape-Maker Spinner	shapespinner.pdf
133	Sharing Shapes Spinner	sharingspinner.pdf
134–136	Sharing Shapes Game Board	sharingboard.pdf
137	Sharing Shapes Sheet	sharingsheet.pdf

Additional Resources	
Title	Filename
CCSS, WIDA, and TESOL	standards.pdf

Notes

Notes

Notes

Notes